A BITE OF GUANGDONG

食在廣東

廣東省人民政府新聞辦公室　編

COMPILED BY THE INFORMATION OFFICE OF THE PEOPLE'S GOVERNMENT OF GUANGDONG PROVINCE

南方日報出版社
NANFANG DAILY PRESS

图书在版编目（CIP）数据

食在广东 / 广东省人民政府新闻办公室编. -- 广州：南方日报出版社，2015.8
ISBN 978-7-5491-1303-3

Ⅰ. ①食… Ⅱ. ①广… Ⅲ. ①饮食－文化－广东省
Ⅳ. ①TS971

中国版本图书馆CIP数据核字(2015)第186313号

Shi Zai Guangdong
食在广东　　广东省人民政府新闻办公室　编

出版发行：南方日报出版社
地　　址：广州市广州大道中289号
电　　话：（020）87373998-8502
经　　销：全国新华书店
印　　刷：广州市岭美彩印有限公司
开　　本：787mm×1092mm　1/16
印　　张：8.75
字　　数：100千字
版　　次：2015年8月第1版
印　　次：2015年8月第1次印刷
定　　价：128.00元

投稿热线：（020）87373998-8503　读者热线：（020）87373998-8502
网址：http://www.nanfangdaily.com.cn/press　http://www.southcn.com/ebook

Editorial Committee of A BITE OF GUANGDONG

《食在广东》编委会

Contents
目录

CANTONESE CUISINE: A BRIEF INTRODUCTION
粤菜探源

CULINARY DELIGHTS IN HISTORIC LINGNAN REGION
寻味岭南

FOOD TRADITIONS
传统食俗

LOCAL SPECIALTIES
缤纷特产

年
好
年
合

CANTONESE CUISINE: A BRIEF INTRODUCTION

粤菜探源

Guangdong Province, located south of China's Nanling Mountain Range on the South China Sea coast, borders Guangxi Zhuang Autonomous Region to the west, Hunan and Jiangxi provinces to the north, Fujian Province to the northeast, and Hong Kong and Macao Special Administrative Regions to the south. The province, which faces Hainan Province across the Qiongzhou Strait, is usually referred to as the "Southern Gateway" to China's foreign trade. It also leads the country in terms of regional gross domestic product (RGDP) and population. The fascinating region, which has a long rich history, vast land areas and a long coastal line, is set to embrace a brand-new future. Guangdong people have always been the bellwether and forerunner in various fields, be it in expanding maritime trade, moving on to modern civilization, or promoting reform and opening up. Guangdong boasts beautiful mountains, rivers and islands, as well as bustling skyscrapers and office buildings in the Pearl River Delta metropolitan area. Ever since the ancient times, Guangdong, which is well-known for its open-mindedness to cultural diversity, has been embracing guests, businessmen and tourists from home and abroad. The rapidly-changing province welcomes all to stay, work and enjoy life here.

Guangzhou, the capital city of Guangdong, is China's third largest city and the center of economy, culture, education and transportation in South China. A nearly 100-year-old popular Chinese saying goes that "Guangzhou is a paradise for eating," crowning Cantonese cuisine one of the most celebrated cuisine in China.

The saying highlights that Guangdong's culinary culture shines with unique glamour thanks to its long history, great diversity and profound influence.

Guangdong food remained obscure for a long time in ancient China. The economy of Lingnan Region (south of the Five Ridges) has seen robust development since the Tang and Song dynasties (618-1279) as China's economic center moved southward and the advanced technologies in the Central Plains in the middle and lower reaches of the Yellow River reached the southern part of the country. Lingnan became the wealthiest region in China during the Qing Dynasty (1644-1911) due to the Canton Trade System (1757-1842) which made Guangzhou China's only port for Western trade. It's a principle of law that business prosperity always leads to booming of catering. Guangzhou emerged as one of the world's largest trading ports and a vibrant international metropolis during the Tang Dynasty. Consequently, the city became a capital of gastronomy. After the five treaty ports were forced to open under the Treaty of Nanking signed in 1842, the Cantonese cuisine was introduced by local businessmen into the northern part of China in cities including Shanghai, Tianjin and Beijing. The unique culinary style gradually rose to fame, especially in Shanghai. During the Republic of China period (1912-1949), Guangdong became the cradle of revolution and fountainhead of new forms of economy and culture. Two Cantonese restaurants--Taishi (Court Historian) Cuisine in Beijing and Tan's Imperial Cuisine in Guangzhou--together opened the trend of "tasting Guangzhou", while in Shanghai, Haipai cuisine-styled Cantonese cuisine (Cantonese cuisine integrated with all-embracing Haipai cuisine, a form of Shanghai cuisine), won the title of "national cuisine" and brought Cantonese cuisine to a peak of development to represent the flavor of the time. The development of Guangdong's culinary culture underwent several ups and downs after the founding of the People's Republic of China in 1949. The Cantonese cuisine restaurants became ubiquitous again throughout the country after the opening-up and reform policy was initiated in late 1970s. Restaurants with Guangdong chefs cooking fresh seafood are regarded as high-end ones in the northern part of China.

People regard food as their primary want, and it's a natural right to be a foodie. Sun Yat-sen, the forerunner of China's revolution to end feudalism, once said, "Delicious food should also be viewed as artworks since good-looking things and pleasant musical notes are both artworks." Guangdong people have created so many wonders to satisfy foodies' passion for delicacies. Gourmet food embodies the most robust part of Guangdong's culture. Fine traits in Guangdong people's disposition, such as pioneering spirit, open-mindedness, inclusiveness and pragmatism, are all reflected in their culinary culture.

Sun Yat-sen represents well the pioneering spirit. Sun, born in a gourmet village in Guangdong, was a gastronome well-versed in culinary art. He invented a famous dish named "Siwu Soup," which is cooked with four ingredients--the Jew's ear, daylily, tofu and bean sprouts. Guangdong was the first to cultivate edible fungi in modern China: Residents in the northern part of Guangdong took the lead in cutting logs into short lengths to cultivate black mushrooms, and monks at the Nanhua Temple in Qujiang County, Shaoguan City learned how to grow straw mushrooms long ago.

Guandong people are open-minded. They sailed across the sea a long time ago and settled down in various countries and regions in the world, including Japan,

广东，位于南岭以南，南海之滨，与香港特区、澳门特区、广西、湖南、江西和福建接壤，与海南隔海相望，是中国对外贸易的“南大门”，也是中国经济总量第一、人口规模第一的省份。这里，既属于悠久的历史，也属于崭新的未来；既属于广阔的大陆，也属于蔚蓝的海洋。这块土地上的广东人，无论开拓海上贸易，开启近代文明，还是推动改革开放，都以“领头羊”的身份“敢为天下先”。广东有美丽的山川和海岛，也有珠三角都市圈的繁华高楼。自古以来，广东就以开放的文化胸襟迎接着海内外的宾客与商旅。今天，日新月异的广东，欢迎着您的莅临。

广州是广东的省会、中国第三大城市，是华南地区经济、文教、交通中心。“食在广州”，是近100年来流传坊间的民谚，赞美粤菜是“舌尖上的中国”的杰出代表。

“食在广州”，表明广东饮食文化源远流长、博大精深、魅力诱人。

广东饮食最早在中国寂寂无名。唐宋以后，随着中国经济大势的南移和中原先进技术的南传，岭南经济得到强劲开发。清末以来，一口通商的优势更使得岭南富甲天下。商业繁荣催生了饮食业的繁荣，这是一条铁律。唐代，广州崛起为世界性贸易大港、国际化大都市。与之匹配，广州同时成为美食之都。五口通商时期，粤菜随粤商北上，到达上海、天津、北京，特别是在上海地区，渐渐获得名声。及至民国时期，广州成为革命策源地和新的经济文化衍生地。北京的太史菜与广州的谭家菜遥相呼应，共开“食在广州”之先河；在上海，海派粤菜赢得“国菜”的殊荣，将“食在广州”

The reason why Guangdong is dubbed "paradise for gourmet food" lies in its rich culinary culture. Throughout history, many eminent writers have been inextricably bound up with Cantonese cuisine. Han Yu, a great litterateur in the Tang Dynasty, was once an official in Chaozhou City. He wrote a number of poems to commend Teochew cuisine (Chaozhou cuisine), which amazes foodies with a full array of sea food. Su Shi, an accomplished litterateur in the Song Dynasty (960-1279), wrote a famous poem on litchi which reads: "Feasting on lychee – 300 a day, I shouldn't mind staying

Europe and the United States. Opening Chinese restaurants is a special skill and means of subsistence for many overseas Chinese, as well as a way to manage their homesickness. Cantonese restaurants are the most popular among overseas Chinese restaurants.

Guangdong's inclusiveness is embodied in Cantonese cuisine's blending the culinary styles of northern and southern China and that of the East and the West. Dishes in the northern part of China, including wonton, Dong Po Pork and West Lake Fish, were introduced into Guangzhou during the Tang and Song dynasties. In the late Qing Dynasty, "Manhan Quanxi," literally the complete feast of Manchu and Han courses, became the most dignified and extravagant banquet in Guangzhou. In the Tang Dynasty, traders from different countries lived in "Fanfang" (foreign residential area) of Guangzhou and introduced exotic foods into the city. Guangzhou's residents would immediately refine the recipes once they found anything useful and as a result, developed new dishes suitable for the local taste buds.

Guangdong people are also pragmatic. In a broad sense, this means that they take measures in line with the local conditions. Tan's Imperial Cuisine and Taishi Cuisine were the two most prestigious cuisines in the early years of the Republic of China. Jiang Kongyin, an established scholar titled "Lord Jiang Taishi," created Taishi Cuisine. He bought a farm in the suburbs of Guangzhou and grew various vegetables and fruits there. He sent his fifth son, who graduated from Yenching University, to study agriculture in Japan. After his son returned home, he opened a bee farm to ensure food materials are sufficient and fresh.

eternally here." Guangdong people can cook a lot of dishes using litchi as an ingredient, such as shelled shrimps with litchi and litchi chicken balls.

Most notably, "eating in Guangdong" implies health preservation and essential elements of life. This is usually embodied in the practice of drinking soups. Housewives, no matter whether they are literate or illiterate, are fond of such topics as "expelling dampness from the body," "adding moisture and clearing the lungs" and "nourishing the body." They know well they should regularly make slow-cooked soups for their family members. There are all kinds of soups, the recipes of which are backed by the time-tested old Yin-Yang and Five Element Theory.

Guangdong, located in the subtropical zone, has a mild climate and abundant rainfall year round. As a result, there are always bounties of fresh vegetables and fruits in all seasons, laying a solid foundation for its being a paradise for gourmet food. Cantonese cuisine is composed of Guangfu, Teochew and Hakka cuisines, due to the distinct culinary customs of the three branches of the Han ethnic group—Guangfu (Guangzhou) people, Chaoshan people and Hakka people. These categories of Cantonese cuisine demonstrate three different styles of life: Guangfu people, who reside in the most fertile Pearl River Delta region and enjoy abundant food materials for cooking; Chaozhou people (in great Chaozhou-Shantou area), who live in the coastal areas of eastern Guangdong, focus on seafood dishes; Hakka people, most of whom live in mountainous areas, highlight raw materials from mountains in cooking.

推向时代的巅峰，表征民国味道。新中国成立之后，广东饮食文化几经沉浮。直到改革开放，粤菜馆再度在全国遍地开花，北方大地都将请到广东大厨烹制生猛海鲜作为高档酒楼的标志。

民以食为先。追求美食是天赋人权。正如孙中山所说，既然悦目的东西是艺术品，悦耳的东西是艺术品，那么悦嘴的东西也应该是艺术品。为了一张嘴，广东人创造了多少人间奇迹！美食体现了广东文化最有生命力的部分。广东民性中的“敢为天下先”“开放”“包容”“务实”等优秀特质，都体现在饮食文化里。

孙中山是“敢为天下先”的代表人物。孙中山出生于美食之乡，爱吃懂吃，创过一道菜，名叫“四物汤”，用料是木耳、黄花菜、豆腐和豆芽。广东是近代培养菌的发源地：粤北人率先把木头截成段培养冬菇，曲江南华寺僧人早就学会培育草菇。

至于“开放”，广东人很早就漂洋过海，足迹遍及日本、欧美等世界各地。开办中餐馆，是大量华人、华侨的看家本领和谋生手段，更是舌头不能割舍的乡愁。海外中餐馆，粤馆执牛耳。

“包容”，则表现为粤菜的南北同炉、中外合璧。唐宋时期，北方的馄饨、东坡肉、西湖鱼等传来广州。晚清，北方的满汉全席成为广州最高贵、最讲排场的筵席。唐代，各国商人聚居广州带来各种异国风味菜。广州人但凡有一点可取的，都马上拿过来，按自己的口味改造一番。

“务实”，大而论之，是因地制宜。民国初年最负盛名的粤菜，一个是谭家菜，一个是太史菜。创制太史菜的太史公在市郊买下一个农场，种植各种蔬菜水果，还派毕业于燕京大学的第五子到日本学习农科，回到农场之后又办蜂场，以保证用料充足新鲜。

“食在广东”的底色，是文化。古往今来，不少文豪与粤菜结下不解之缘。唐代的大文豪韩愈在潮州做官时就作诗盛赞潮州菜，海鲜食材林林总总，令人惊叹。宋代的大文豪苏东坡在惠州留下“日啖荔枝三百颗，不辞长作岭南人”的佳句，而广东人用荔枝可以做出荔枝虾仁、荔枝鸡球等好多道菜。

为人称道的是，“食在广东”蕴含着保健意识，通常体现在喝汤这件事上。家庭主妇无论有文化没文化，都爱谈论“祛湿”“清润”“滋补”，都懂得定期煲汤给家人喝。汤的种类不胜枚举，背后都有最古老的阴阳五行理论作为支撑，经历过千锤百炼。

广东地处亚热带，气候温暖，雨量充沛，瓜果菜蔬四时常绿，为“食在广东”奠定了基础。而广府、潮汕、客家三大民系独具一格的饮食习惯，使粤菜分为广府菜、潮州菜、客家菜。这三个流派，呈现出三种生活形态：广府人占据着最肥沃的珠江三角洲，做菜选材丰富；潮州（大潮汕）人占据着东部沿海，做菜以海产品为主；客家人多居住在山区，落入林莽，便以山野材料入菜。

GUANGZHOU CUISINE,

The Most Typical Cantonese Cuisine

广府菜

粤菜的代表

The three categories of Cantonese cuisine feature different characteristics.

Guangfu cuisine was renamed Guangzhou cuisine in the 1960s and 1970s. Originating in Guangzhou, it is the most typical Cantonese cuisine.

Guangfu cuisine is like transformers in animated films that can change themselves into different objects in a dazzling way.

Guangfu cuisine incorporates Nanhai, Panyu, Dongguan, Shunde and Zhongshan cuisines, featuring a combination of relatively light, fresh, clear and fragrant flavors. In this culinary style, dishes are cooked in different ways in different seasons, with light flavors served in summer and autumn and mellow flavors in winter and spring.

Using an extensive variety of ingredients and retaining fresh taste are the two greatest features of Guangfu cuisine. Other distinctive features are as follows: proportions of ingredients are meticulous, condiments are varied and ingenious, and decorations of food are beautiful and brightly colored. The culinary style subtly blends the essence of cuisines in different cities and regions, including Beijing, Suzhou, Yangzhou and Hangzhou, as well as the Western cooking techniques. Chefs are good at innovation based on imitation, integrating different culinary styles to form their special features and thus bringing about more varieties of dishes. Early in 1965, a total of 5,457 kinds of dishes in Guangfu cuisine were showcased at the Guangzhou Famous Dishes and Delicious Desserts Expo. Guangfu cuisine also features upscale dishes, such as the exquisite banquet of edible bird's nest and shark's fin, but there are more family-style dishes, such as white gourd soup and long-simmered soup.

The cooking techniques include stir-frying, frying, sautéing, braising, soft-frying, stewing, boiling, steaming, pickling, spicing, and curing with salt or sugar. The most typical dishes include white boiled shrimp, roast suckling pig, braised sliced pork with taro, Huangpu scrambled egg and stewed tylorrhynchus.

粤菜的三个流派，各具特色，个性鲜明。

广府菜在20世纪六七十年代后改称广州菜，是粤菜的代表，发祥地广州，流行于讲粤语的地方。

广府菜就像动画片中的变形金刚，能够变得令人眼花缭乱。

南海菜、番禺菜、东莞菜、顺德菜、中山菜的特色风味，广府菜将其变成一体，口味比较清淡，力求清中求鲜、淡中求美。而且广府菜的各种菜式随季节时令的变化而变化，夏秋偏重清淡，冬春偏重浓郁。

广府菜的最大特点是用料广、味道鲜。另一突出特点是，用量精而细，配料多而巧，装饰美而艳，且兼京、苏、扬、杭等外省菜以及西菜之所长，善于在模仿中创新，融为一体，自成一家，故而品种繁多。早在1965年，广州名菜美点展览会介绍的广府菜，就多达5 457种。广府菜还有一大特点是，既有高档菜，如最精绝的燕翅席，更多的却是家常菜，如冬瓜盅、老火汤等。

广府菜的烹调方法，有煎、炒、爆、烧、炙、煲、煮、蒸、甑、腌、卤、腊、蜜饯等。代表性菜肴有白灼虾、烤乳猪、香芋扣肉、黄埔炒蛋、炖禾虫等。

TEOCHEW CUISINE,

China's Highest-end Cuisine

潮州菜

中国最高端菜系

Teochew cuisine, a counterpart of Guangfu cuisine, is usually regarded as the top cuisine in China. It is dubbed "China's highest-end cuisine."

The great Chaozhou-Shantou area, located in the subtropical zone, faces the sea to its south with abundant marine products. As the Chinese saying goes that "Those living near the sea live off the sea," Teochew cuisine is well-known for its seafood cooking techniques. Selection of ingredients is meticulous and curing procedures are elaborate. Seafood dishes, served with various fragrant sauces, are fresh and delicious, refreshing but not light, fragrant without fishy smell, and mellow but not greasy. Some of the signature dishes include Mandarin Duck Crab Paste, lettuce lobster, stewed shark's fin, Oyster omelette, boiled eels and consommé crab balls.

Teochew cuisine focuses on both meat and vegetable dishes. Chefs are adept at cooking dishes with vegetables and fruits. Raw materials are cooked with delicate skills to create fresh, refreshing, delicious and nutritious dishes. Teochew dishes commended by foodies include the "Protect Country Dish" favored by Zhao Bing, the last emperor of the Southern Song Dynasty (1127-1279), smashed water chestnut, braised mustard greens and sweet potato stewed with sugar.

Teochew cuisine is a kaleidoscope of dishes and flavors---Apart from seafood and assorted spiced meat, there are more than 100 varieties of cold dishes, such as pickles, kale borecole and dried radish, as well as over 100 kinds of meat balls, with the most famous ones being beef meat balls and fish balls. Teochew-style hot pot and light soup are also well-known.

作为广府菜的兄弟，潮州（大潮汕）菜往往被视为顶级菜，有“中国最高端菜系”之称。

潮州（大潮汕）地处亚热带，南临大海，海产丰富，“靠海吃海”。潮州菜自然最以烹制海鲜见长：选料考究，制作精细，以酱碟佐料，新鲜美味，清而不淡，鲜而不腥，郁而不腻。潮州菜中，海鲜类的代表名作有鸳鸯膏蟹、生菜龙虾、红炖鱼翅、蚝烙、清炖乌耳鳗、清汤蟹丸等。

潮州菜爱荤而不嫌素，善于烹制以蔬果为原料的素菜，是其一大特点：粗料细做，清淡鲜美，营养丰富。为南宋皇帝所喜爱的护国菜，以及马蹄泥、厚菇芥菜、糖烧地瓜等，皆为美食家所称道。

菜式仿如万花筒，是潮州菜的又一大特点：海鲜、卤味之外，还有咸菜、橄榄菜、菜脯等100余种杂咸，以及100余种丸子，最出名的属牛肉丸和鱼旦。潮式火锅、淡汤也是榜上有名的。

潮州菜之所以享有盛名，还在于刀工讲究，制作精妙，加工方式依原料特点而多样化，有煎、炒、烹、炸、焖、炖、烤、焗、卤、熏扣、泡、滚、拌等。此外，厨师用竹笋、萝卜、薯类，精工雕刻成各式各样花鸟造型，作为点缀或菜垫，形成独特的“彩盘”艺术。

潮州菜的筵席自成一格，大席喜用12道菜，其中咸、甜点心各一道。喜席必有头尾两道甜菜，称为“头尾甜”，让客人吃得可口，心情也好。

Teochew cuisine enjoys a high reputation, partly because the cutting techniques are exquisite and curing procedures are elaborate. The processing methods vary according to the traits of raw materials. Cooking skills are varied, including quick-frying, sautéing, frying and stirring in prepared sauce, deep-frying, stewing, braising, roasting, baking, spicing, marinating, smoking and coating, soaking, quick boiling and stirring. The plate dressing style in Teochew cuisine is colorful and unique, as chefs would decorate dishes with flowers and birds carved from bamboo shoots, radish or potatoes.

Teochew banquets are also distinctive. A big full banquet usually offers a feast of 12 courses, with two dimsums, one sweet and one salty. There must be two desserts for a wedding banquet, indicating "being sweet from the beginning to the end." The arrangement creates a more delicious feast for guests and gives them a good mood.

HAKKA CUISINE,

Gourmet Food in Mountains

客家菜

大山里的美食

If Teochew cuisine is compared to South China Sea girls who are pure and elegant, Hakka cuisines is like unsophisticated and steady girls in mountainous areas.

The ancestors of Hakka people left the Central Plains and made an arduous journey to settle down mostly in the mountainous areas of eastern Guangdong. Where there are mountains, there are Hakka people. As mountain dwellers, they live off the mountains. Hakka cuisine features the following six characteristics:

Salty: Hakka dishes are usually a little salty; Fat: Hakka dishes are often soaked in oil and soup as a lot of oil is used in cooking; Fragrant: The most-employed cooking methods are frying, deep-frying, sautéing and braising, in order to enhance the aroma of a dish; Hot: Dishes should be served when they are hot, and "being hot" is used as a criteria to judge whether a dish is delicious or not; Well-done: Dishes should be thoroughly cooked and shouldn't be served raw or half-cooked; Preserved: Pickled and preserved vegetables are popular since fresh ones would deteriorate during a long journey.

Hakka cuisine has the following four primary features: First, Meat of domestic fowl and livestock, as well as of wild game, is the most-used ingredient. There is a saying that "Soup without chicken cannot be clear, food without meat cannot be fragrant, dining without duck cannot be refreshing, and juice without pork leg cannot be mellow." Second, the main ingredient is the centerpiece in the curing process. The main food material is usually cut into chunks, and cooked in a way to be soft, fragrant and mellow. Thick or strong-flavored seasonings are rarely or never added. Instead, raw scallion and cooked garlic are the commonly used condiments. Third, timing is given top priority in Hakka cuisine, and the best cooking techniques include stewing, toasting, simmering and braising. Hakka clay-pot casseroles are extremely famous. Fourth, the appearance of Hakka dishes is traditional-styled and unsophisticated, with strong authentic local flavors.

The traditional signature Hakka dishes include braised sliced pork with preserved vegetables, Hakka salt-baked chicken, Hakka stuffed tofu cubes (with minced meat), pork stomach wrapped chicken, stuffed bitter melon, Poon Choy (Pot with assorted delicacies), Four Stars Longing for the Moon (Sliced carp steamed with scallions in a spicy sauce, surrounded by a quartet of smaller dishes), Hakka dumpling and Hakka steamed stuffed bun. Different from Guangfu Cuisine and Teochew cuisine which underwent significant changes in the fusion of the Han and Yue cultures after the Qin Dynasty (221 BC-207 BC), Hakka cuisine has retained the traditions in the Central Plains, and some of its cooking techniques were inherited from the ancient practices and have evolved to thrive independently.

如果说，潮州菜有南海姑娘的清纯淡雅，那么，客家菜就有山妹子的纯朴稳重。

客家人从中原跋山涉水而来，多居粤东山区，"逢山必有客"，故而"靠山吃山"。客家菜特色口味唱的是"六字山歌"——咸：吃在客家，咸是一绝；肥：用油很足，以至于"半油半汤"；香：多煎、炸、炒、焖，但求"香口"；烧：菜要趁热吃，"烧不烧"就是好不好吃；熟：讲究熟透，忌生或半生不熟；陈：长途跋涉，新鲜蔬菜不好带，干菜、腌菜自然受青睐。

客家菜的基本特色，一是用料以肉类为主，取自家养禽畜和山间野味，有"无鸡不清，无肉不鲜，无鸭不香，无肘不浓"的说法；二是加工突出主料，讲究粗刀大块，讲求酥软香浓，极少添加甚至不加过重过浓的佐料，一般用生葱熟蒜调味；三是技艺注重火功，以炖、烤、煲、酿见长，尤以砂锅菜闻名；四是菜式造型古朴，乡土风貌明显。

传统的客家招牌菜，有梅菜扣肉、盐焗鸡、客家酿豆腐、猪肚包鸡、酿苦瓜、盆菜、四星望月、芋子包、芋子饺等。与广府菜、潮州菜曾经历秦代以后的"汉越融合"阶段不同，客家菜一直保留着中原传统，有的烹饪技艺是古法的继承和演变。

CULINARY DELIGHTS IN HISTORIC LINGNAN REGION

寻味岭南

PEARL RIVER DELTA REGION

珠三角

The Pearl River Delta, or simply "Zhusanjiao" for short, is formed by three major rivers – the Xijiang River (West River), the Beijiang River (North River) and the Dongjiang River (East River) – flowing into the South China Sea. Covering an area of about 56,000 square kilometers in south central Guangdong Province, it is composed of nine cities, namely Guangzhou, Shenzhen, Foshan, Dongguan, Zhongshan, Zhuhai, Huizhou, Jiangmen and Zhaoqing. Located at the lower reaches of the Pearl River, the region borders Hong Kong and Macao with Southeast Asia across the sea. Due to its convenient land and sea transportation, it is known as the "Southern Gateway" of China.

Since ancient times, the Pearl River Delta has embraced a prosperous economy. With a brisk trade and the booming culture and education industry, it has always taken the lead in Guangdong Province, even in south China. The agricultural and sideline products and handicrafts here enjoy a vast market, and their production and sales are quite impressive. Nowadays, the Pearl River Delta region is worldwide influential as an advanced manufacturing industry base and a modern service industry base. As the main area for the country to participate in economic globalization, it is the national base for science and technology innovation and technology research and development; as an important engine for national economic development, it takes the lead in driving the development of southern, central and southwestern areas of China. It is also one of the three regions in China with the largest population, the most innovative ability and the highest general competence.

Local residents of the Pearl River Delta region are largely the Cantonese people, so Guangzhou cuisine is representative here.

珠江三角洲，简称珠三角，是西江、北江和东江入海时冲击沉淀而成的一个三角洲，面积大约5.6万平方公里，位于广东省中南部，珠江流域下游，包括广州、深圳、佛山、东莞、中山、珠海、惠州、江门、肇庆共9个市。毗邻港澳，与东南亚地区隔海相望，海陆交通便利，被称为中国的“南大门”。

珠江三角洲自古经济发达，贸易繁荣，文教鼎盛，一直处于广东省，乃至南方地区的领先地位，农副产品和手工业产品市场广阔，产销活跃。当今，珠三角地区是有全球影响力的先进制造业基地和现代服务业基地，我国参与经济全球化的主体区域，全国科技创新与技术研发基地，全国经济发展的重要引擎，辐射带动华南、华中和西南地区发展的龙头，我国人口集聚最多、创新能力最强、综合实力最强的三大区域之一。

珠三角本地居民以广府民系为主，广府菜是珠三角地区的代表性菜式。

GUANGZHOU

A World-renowned Gourmet City

广州

食在广州，名扬天下

Located in south central Guangdong Province, Guangzhou City is composed of twelve districts, namely Yuexiu District, Haizhu District, Liwan District, Tianhe District, Baiyun District, Huangpu District, Huadu District, Panyu District, Nansha District, Luogang District, Conghua District and Zengcheng District. Covering an area of 7,434.4 square kilometers, the city is the national advanced manufacturing industry and modern service industry base, the national flower base as well as the City of Fruit. Major tourist attractions here include Baiyun Mountain, the night scenery of the Pearl River, the Chen Clan Ancestral Hall, Huanghuagang Park, the Guangdong Olympic Sports Center, Lotus Mountain, the Sun Yat-sen Memorial Hall, the Huangpu Military Academy, the Museum of the Mausoleum of the Nanyue King, the South China Botanical Garden, Conghua Hot Spring, Baomo Garden, the Guangdong Museum of Art, the Shangxiajiu Pedestrian Street, the Beijing Road Pedestrian Street and the Canton Tower. There are local specialities such as Zengcheng Simiao rice, Gualyu lychees and Luogang sweet oranges.

Guangzhou is one of the first batches of national famous historical and cultural cities. As a legend goes, in ancient times, there were five celestial beings riding on five different colored rams to Guangzhou, each holding an ear of rice in its mouth. They presented these ears of rice as a gift to local people and blessed them with “no famine forever.” Hence, Guangzhou is also known as the ram city and the Sui city (Rice-Ear city). Nowadays, the five-ram statue in Yuexiu Park has become the symbol of Guangzhou. Besides, Guangzhou is also a heroic city with a glorious revolutionary tradition. In the modern history of China, there was Sanyuanli Anti-British Struggle to fight against imperialist aggression, the "3.29" Guangzhou Uprising (also known as the Huanghuagang Uprising) led by Sun Yat-sen to fight against feudal rule, and the Guangzhou Uprising launched by the Communist Party of China.

广州市位于广东省中南部，辖越秀区、海珠区、荔湾区、天河区、白云区、黄埔区、花都区、番禺区、南沙区、从化区、增城区。行政区域面积7 434.4平方公里，是全国先进制造业和现代服务业基地、全国花卉基地、水果之乡；主要旅游景点有白云山、珠江夜景、陈家祠、黄花岗、广东奥林匹克中心、莲花山、中山纪念堂、黄埔军校、南越王墓博物馆、华南植物园、从化温泉、宝墨园、广东美术馆、上下九路商业步行街、北京路商业步行街、广州塔等；土特产有增城丝苗米、挂绿荔枝、萝岗甜橙等。

广州是全国第一批历史文化名城之一。传说古代有5位仙人，骑五色羊，羊衔谷穗，降临广州，把谷穗赠予百姓，祝愿这里“永无饥荒”。因此广州又称为羊城、穗城。如今，越秀公园的五羊雕像已成为广州的象征。广州也是具有光荣革命传统的英雄城市。在近代史上有三元里人民反抗帝国主义侵略的抗英斗争、孙中山领导的反对封建统治的“三二九”起义(又称黄花岗起义)、中国共产党领导的广州起义。

▸ Link: Recommended restaurants

Guangzhou Restaurant

The Guangzhou Restaurant, a well-known Chinese brand, earns its name from the meaning of "dining in Guangzhou" and has a history of 80 years. Headquartered at No. 2 Wenchang South Road of Guangzhou, the restaurant was founded in 1935 and formally known as the Southwest Restaurant, which was destroyed by fire in 1938. Two years after the fire, the restaurant was reconstructed by Chen Xinghai, Guan Lemin and Liao Bi and was renamed the Guangzhou Restaurant. Funded by Cai Weihan, it reopened after a long-time closing since February 1950. It became a state-owned restaurant in February 1956. From the end of the 1950s to the end of the 1970s, due to a scarcity of food materials, the Guangzhou Restaurant used vegetables as ingredients to make dim sum to ensure supplies. It also improved cooking techniques to exhibit innovative cuisine at the Canton Fair.

The Guangzhou Restaurant has grown rapidly since the reform and opening-up policy was implemented. It was rated as a national grade-II enterprise by the Ministry of Commerce in September 1988. Since it opened the first branch on December 30, 1989, it has expanded to the locations in Bingjiang West, Tianhe Stadium East, Huangpu, Wuyang New City and Fang Village.

The Guangzhou Restaurant was named one of "China's Top 10 Brand Enterprises" by the Ministry of Commerce, and won the titles of "China's Time-Honored Brand," "National Enterprise of Customer Satisfaction" and "China's Well-known Brand." Dedicated to exploring traditional Cantonese culinary culture and renovating dishes, the restaurant has won acclaim for making ancient style dishes.

广州酒家

广州酒家是"中国驰名商标"，依"食在广州"之意而名，至今已有80年历史。广州酒家总店位于广州市文昌南路2号，始建于1935年，前身为西南酒家。两年后，陈星海、关乐民、廖弼等人集股重建复业，更名为广州大酒家。改革开放后，广州酒家发展迅速。1988年9月，被商业部评定为国家二级企业。1989年12月30日起，在海珠区开设了滨江西广州酒家分店，现有滨江西、天河体育东、黄埔、五羊新城、芳村等多家分店。

广州酒家被国家商务部评为"中国十大餐饮品牌企业"，先后获"中华老字号""全国用户满意服务企业""中国驰名商标"等荣誉，致力于发掘传统粤菜文化和创新菜品服务，因先后推出"满汉大全筵""五朝华筵"和"南越王宴"等仿古宴席而备受赞誉。

White Swan Hotel

The Guangzhou White Swan Hotel is China's first five-star hotel built in cooperation with a foreign company and is also China's first large-scale and modern hotel designed, constructed and managed by Chinese people. It opened in 1983 and was admitted by the Leading Hotels of the World as the first member from China in 1985. The hotel has been named the first-choice hotel for business travelers by international travel guide, magazines and newspapers for many consecutive years.

Since the opening, the hotel has received over 40 heads of state and government, including Queen Elizabeth II of the United Kingdom, President Bush of the United States, President Nixon of the United States, Chancellor Kohl of Germany, Castro, Kissinger, Sihanouk and Lee Kuan Yew.

Deng Xiaoping, the "chief architect" of Chinese reforms, visited the hotel for three times and wrote an inscription.

白天鹅宾馆

广州白天鹅宾馆是中国第一家中外合作的五星级宾馆，也是中国第一家由中国人自行设计、 施工、管理的大型现代化酒店。1983年开业，1985年被世界一流酒店组织接纳为在中国的首家成员。连续多年被国际旅游指南和国际著名报刊评为国际商务人士到广州的首选酒店。

开业以来，白天鹅宾馆共接待了40多个国家的元首和政府首脑，英国女王伊丽莎白二世、美国前总统老布什及尼克松、德国前总理科尔、古巴前主席卡斯特罗、美国前国务卿基辛格、柬埔寨前国王西哈努克、新加坡前总理李光耀等国际名人都曾在此驻足。中国改革开放的总设计师邓小平更是三次莅临，并亲笔题字。

Garden Hotel

The Guangzhou Garden Hotel is currently China's largest five-star business hotel. Initiated by Liao Chengzhi and Li Mingze in 1982, the hotel is jointly operated by Guangzhou Lingnan Property Corporation and Hong Kong Garden Hotel Co., Ltd. It was established in October 1984. Over the past three decades since the opening in 1985, the internationally renowned hotel has been best known for its strong historical and cultural appeal as well as outstanding services. The logo of the hotel is bombax ceiba, the official flower of Guangzhou.

The Guangzhou Garden Hotel is one of China's only three platinum five-star hotels authorized by the National Tourism Administration and is the only Chinese brand hotel that gets authorized in South China. The Y-shaped hotel comprises two buildings of Hotel Tower and Garden Tower and occupies a site area of 48,000 square meters, a construction area of 170,000 square meters and a total garden area of 21,000 square meters. Standing at 107 meters high, the 30-story hotel has a revolving restaurant atop. It has more than 2,000 rooms, suites, apartments and offices, as well as 14 restaurants and bars, providing a wide-ranging menu of flavors from China, France, Italy and Japan as well as diversified quality services to guests who have varied needs.

Dong Fang Hotel

The Dong Fang Hotel is the only garden-style hotel with sub-tropical features in the heart of Guangzhou. At the end of 2005, the hotel received the International Star Diamond Award from the American Academy of Hospitality Sciences. The award is regarded as the most prestigious award recognizing excellent services in finance, hospitality, aviation and gastronomy sectors.

Having a history of 50 years, the Dong Fang Hotel is located in the city's busy commercial hub—Liuhua Commercial District, and is adjacent to the famous shopping and recreation centers as well as scenic spots, such as Beijing Road Pedestrian Street, Yuexiu Park and the Museum of the Western Han Dynasty Mausoleum of the Nanyue King. There are a lot of delicacies, recreational facilities and shops nearby for travelers to explore.

花园酒店

广州花园酒店是目前国内最具规模的五星级商务酒店。1982年由廖承志、利铭泽倡议兴建，广州岭南置业公司与香港花园酒店有限公司合作经营。1984年10月建成。从1985年开业至今，以丰富历史文化底蕴和卓越服务著称，享誉中外30年。其店徽为广州市花木棉。

广州花园酒店现为中国国家旅游局正式授予的白金五星级饭店（目前全国仅有三家），是华南地区唯一获选的中国自主品牌酒店。酒店分酒店大楼和花园大厦两座Y形大楼，占地面积4.8万平方米，建筑面积17万平方米，其中花园占地2.1万平方米。酒店大楼有30层，高107米，顶楼设旋转餐厅，拥有2 000多间客房、套房以及公寓和写字楼，并有14家餐厅及酒吧，荟萃中、法、意、日等多国美食，为不同需要的宾客提供高水准的多元化服务。

东方宾馆

东方宾馆是广州城市中心唯一的亚热带风情绿色园林酒店，2005年末，荣获全球服务领域的最高奖项“国际五星钻石奖”（该奖由美国优质服务科学学会评选，是国际公认最具权威的优质服务奖，主要颁发给在金融业、酒店业、航空业和餐饮业等服务领域做出突出贡献的企业和个人）。

东方宾馆已有50年历史，坐落在繁华的城市商业中心——流花商圈，毗邻北京路步行街、越秀公园、西汉南越王墓博物馆等著名购物娱乐中心和旅游景点。周边拥有丰富多彩的美食、娱乐和购物选择。

SHENZHEN

A Fast-growing and Inclusive Metropolis

深圳

新兴城市，新旧包容

Shenzhen is located in southern Guangdong Province. It was established as a prefecture-level city in 1979 and elevated to a sub-provincial city in 1981, It has jurisdiction over six administrative districts (Futian, Luohu, Nanshan, Yantian, Baoan and Longgang) and four function areas (Guangming New Area, Longhua New Area, Pingshan New Area and Dapeng New Area), covering an administrative area of 1,997 square kilometers. Its main tourist attractions include Happy Valley, the Window of the World, Splendid China, China Folk Culture Village, Minsk World, Dameisha, Xiaomeisha and Dapeng Fortress.

Shenzhen is abundant in aquatic resources, as over 40 species of rare fish are found here, including brown-striped mackerel scad, round sardinella, spotted scat, macracanthus, hairtail, porgy, silver sea perch, bass and shrimp, as well as crab, shellfish and algae. The local delicacies include Shajing oyster, Nanshan lichee, Xili mango, Dapeng cloud-fog tea, Gongming roast goose, Xixiang greasyback shrimp, Shiyan pear, Nan'ao abalone, Nan'ao sea urchin, Pingshan orange, Yongfu mullet, Songgang preserved duck, Longgang sanhuang chicken, Guangming pigeon and Nanshan peach.

The native people in Shenzhen are simple and honest and have well kept folk custom. Earthen building is a traditional residence for Hakkas. Poon choi is a traditional cuisine commonly served in Futian's Xiasha and Nantou. Mazu Birthday Festival in each year's March is a temple fair held in the Tianhou Temple by fishermen and farmers in coastal Shenzhen to pray for safe navigation, prosperous fishing and harvest.

深圳市位于广东省南部。1979年设立地级市，1981年升格为副省级市，辖福田区、罗湖区、南山区、盐田区、宝安区、龙岗区等6个行政区，光明新区、龙华新区、坪山新区、大鹏新区等4个功能区。行政区域面积1 996.78平方公里。主要旅游景点有欢乐谷、世界之窗、锦绣中华·民俗文化村、明思克航母世界、大小梅沙和大鹏所城等。

深圳市水产资源丰富，有蛇鲻、蓝圆鲹、金色小沙丁、金钱鱼、大眼鲷、带鱼、三刺鲷、盲曹和鲈鱼等40余种名贵鱼种，还有虾、蟹、贝类和藻类。土特产有沙井鲜蚝、南山荔枝、西丽芒果、大鹏云雾茶、公明烧鹅、西乡基围虾、石岩沙梨、南澳鲍鱼、南澳海胆、坪山金龟橘、福永乌头鱼、松岗腊鸭、龙岗三黄鸡、光明乳鸽、南山甜桃等。

深圳原居民民风淳朴，保留较多的民间习俗。客家围龙屋为典型客家民居建筑；大盆菜是福田下沙和南头一带的习俗；妈祖诞庆是每年3月深圳沿海渔民和农民在天后庙举行的庆祝庙会，祭祀天后，求航海平安、渔业兴旺、五谷丰登。

▸ Link: Recommended restaurants

Shenzhen Venice Hotel

The Shenzhen Venice Hotel is located in the picturesque and lakeside Overseas Chinese Town (OCT) and is China's first international five-star business resort featuring Venice culture. Located in southwestern Shenzhen, the hotel is adjacent to the famous theme parks such as the Window of the World, the Splendid China Folk Culture Village and China Folk Culture Village. The hotel enjoys a convenient transportation, only several minutes from the West Passage, Futian Center District and Science & Technology Park, 15 minutes from Shekou Wharf and Huanggang Port, and 20 minutes from Luohu Port.

Because of its unique theme culture and excellent services, the Shenzhen Venice Hotel has received wide recognition and attention from the public and media. Over the past few years, the hotel has won many awards, such as the "Best Business Hotel in Shenzhen" by Business Traveler Asia-Pacific, "China' s Best Business Hotel" by Fortunes Chinese. The hotel offers a romantic rooftop swimming pool, exciting bar and stylish Italian restaurant, alluring the city's celebrities to come.

Laurel Restaurant, OCT

The Laurel Restaurant is a famous restaurant in Shenzhen. Its OTC store is located in a villa complex on Xiangshan Street, overseeing a lake view that won't disappoint any customers. It is noted for Cantonese cuisine, with delicious sliced duck, savory pork trotter and poplar din sum on offer.

Victory Restaurant, Lizhi Park

Located at the east gate of Lychee Park on Hongling Central Road, Futian District, the restaurant is the first branch of Victory Restaurant. Complemented by the neighboring park's scenery, the restaurant is designed with primitive garden style. The restaurant has outdoor dining tables with umbrellas and swings, while its indoor dining space is a bit crowded. When it comes to dishes, the restaurant offers impeccable tastes, such as succulent clam with black bean sauce, special fried herb jelly, simmered fish head, crab vermicelli pot and durian custard pie. It also serves up fine tea and late snack.

深圳威尼斯酒店

深圳威尼斯酒店位于深圳湾畔风光秀丽的华侨城，是中国首座以威尼斯文化为主题的国际品牌五星标准商务度假酒店。酒店位于深圳的西南部，紧临著名的主题公园世界之窗、锦绣中华·民俗文化村。交通便利，酒店距西部通道、福田中心区和科技园仅数分钟车程，距蛇口码头、皇岗口岸仅15分钟车程，而且距罗湖口岸仅20分钟车程。

深圳威尼斯酒店因其独特的主题文化和周到热情的服务，而备受公众和媒体的认可和关注。在过去的几年中，酒店获得了不同媒体的多种奖项，如由亚太《商旅杂志》评选的“深圳最佳商务酒店”、由《福布斯》杂志中文版评选的“中国最优商务酒店”等。深圳威尼斯酒店浪漫的空中泳池、流行劲爆的V吧、优雅的蓝色意大利餐厅都是城中名流的“会聚之所”。

丹桂轩华侨城店

丹桂轩是深圳的知名酒家，这家分店坐落在南山区华侨城香山街波托菲洛水岸别墅区，有景观位可供人临湖远眺，环境绝不会让你失望。这里的粤菜以精致著称，片皮鸭和猪手可圈可点，点心也相当受欢迎。

胜记荔枝公园店

胜记荔枝公园店是本土酒家胜记的第一家店，位于福田区红岭中路荔枝公园东门，依着公园的好景致，古朴的园林风很是搭调，露天位有阳伞和秋千。经典菜式豉汁黄沙蚬汁香味浓郁，炸龟苓膏是一大特色，啫啫鱼头、膏蟹粉丝煲和榴莲酥等都很精致。

ZHUHAI

A City of Hundred Islands and Capital of Seafood

珠海

百岛之市，海鲜之城

Zhuhai City is located in the southern part of Guangdong Province. Set up in 1979 as a prefectural city, it consists of the districts of Xiangzhou, Jinwan, Doumen and has established Zhuhai High Technology Industrial Development Area, Zhuhai Bonded Area, Gaolan Port Economic Zone, and Wanshan Ocean Development Pilot Zone. The city's land and sea area totals 7653 square kilometers, of which land area amounts to 1724.32 square kilometers.

With the largest sea and island area, the biggest number of islands and the longest coastline among cities in the Pearl River Delta region, it is known as the "City of Hundred Islands" for its 217 islands of different sizes and its shoreline stretching as far as 224.5 kilometers. Gaolan Port is one of China's main coastal hubs, Gongbei Port is the nation's second largest importing and exporting land port, while Jiuzhou Port is the largest importing and exporting sea port in China.

Among its various specialties are oyster sauce, Zhuhai fatty crabs, Huangjin finless eels, Hengshan pueraria thomsonii benth, Baiteng lotus root starch, teals, Qianwu daces, Nanping crisped grass carps, Baijiao tylorrhynchus, grassworks, Xiaotuoshan oranges, Huangyang lychees, and Wanzai flowers.

The city's main tourist destinations and sceneries include Golden Beach, Pearl Land, Statue of Fishing Girl, New Yuanming Palace, Chimelong Oceankingdom, Huangyang Mountain, Ocean Spring Resort, Martyrs' Cemetery, Tangjia Gongle Park, White Stone Street, Dong'ao Island, Zhuhai Imperial Hot Spring Resort, Stone Park, the Outer Lingding Island, Guishan Island, Jiuzhou Island, Shijing Hill, Lost City Water Park, Lata stone fort, Dangan Island Nature Reserve. It is also home to national key cultural relics protection units, namely Baojing Bay site and Chenfang Homesteads.

珠海市位于广东省南部。1979年设立地级市，辖香洲区、金湾区、斗门区，另设横琴新区、珠海高新技术产业开发区、珠海保税区、高栏港经济区、万山海洋开发试验区。全市海陆域总面积7 653平方公里，其中陆地面积1 724.32平方公里。珠海是珠三角城市中海洋面积和海岛面积最大、岛屿最多、海岸线最长的城市，素有“百岛之市”之称。全市大陆海岸线224.5公里，大小岛屿217个。高栏港是中国沿海主枢纽港，拱北口岸是中国第二大陆地进出境口岸，九洲口岸是中国第一大海港进出境口岸。

主要特产有叠石蚝油、珠海肥蟹、黄金风鳝、横山粉葛、白藤藕粉、水鸭、乾务软骨鲮、南屏脆肉鲩、白蕉禾虫、草织品、小托山橘、黄杨荔枝、湾仔鲜花等。主要旅游景点有金海滩、珍珠乐园、珠海渔女、圆明新园、长隆海洋王国、黄杨山、海泉湾、烈士陵园、唐家共乐园、白石街、东澳岛、珠海御温泉度假村、石博园、外伶仃岛、桂山岛、九洲岛、石景山、珠海梦幻水城、珍珠乐园、拉塔石炮台、担杆岛自然保护区。全国重点文物保护单位有宝镜湾遗址和陈芳家宅。

▸ Link: Recommended restaurants

Zhuhai Grand Bay View Hotel

Zhuhai Grand Bay View Hotel is situated at the South Qinglv Road, Gongbei Community, Xiangzhou District. Sitting against the luxuriant General Hill and facing the immerse South China Sea, the hotel is just five-minute drive away from Macau gateway and Jiuzhou Hong Kong-Shenzhen Aerofoil Boat Dock, making it easy to arrive in Hong Kong, Guangzhou and Shenzhen within 90 minutes.

On the ground floor of the hotel, you can find a customer service center, a waterfall lounge with a panoramic view of artificial rockeries, a shopping center full of world-known brands, and Isabella Beauty Salon.

On the second floor, there is an antique Chinese dining room in Tang style, a cafeteria having an open kitchen where live cooking performances are staged, a transparent wine cellar showcasing the world's famous wines, a dining area providing authentic northern wheaten food, the well-equipped Yuanda banquet hall and a conference hall.

The Fifth to 14th floors are customer-friendly guest rooms with a safe, hygienic and cozy environment.

珠海海湾大酒店

珠海海湾大酒店位于珠海拱北情侣南路，背靠苍翠欲滴的将军山，面临碧波浩渺的南中国海；距澳门关口和九洲港港深飞翼船码头仅5分钟车程，90分钟内即可抵达香港、广州及深圳等地，交通十分便利。珠海海湾大酒店一层大堂设有温馨、周到的客户服务中心，拥有全景假山的瀑布酒廊、汇聚世界名牌的购物商场，以及伊莎贝拉美容美发中心。二层设有古风古韵的唐宫中餐厅、拥有开放式厨房及现场厨师表演的咖啡西餐厅、珍藏各国名牌葡萄酒的通透式酒窖、拱形地道的北方面食及远大宴会厅、会议厅。珠海海湾大酒店5—14层为客房区域，人性化的空间规划与房间设计，为您提供安全、卫生、舒适的休憩环境。

Zhuhai Yuecai Holiday Inn

Zhuhai Yuecai Holiday Inn is a 30-storied modern hotel in the bustling central business district of Zhuhai. It's just 15-minute drive away from the Gongbei Port square, 10 minutes away from Jiuzhou Port Dock where ships commuting between Hong Kong, Shenzhen Shekou ports and Zhuhai, and 45 minutes away from Zhuhai Airport.

The hotel has launched free shuttle buses to Jiuzhou Port Dock since November 2001. At the Caihuaxuan Chinese dining hall, customers can enjoy Cantonese dishes prepared by celebrity chefs, new Hong Kong-style desserts and various famous dishes. Lucio Italian Restaurant has an open kitchen with a brick pizza oven. At Blarney Stone Pub, the first Irish pub in Zhuhai, spirits lovers can have a sip of Irish flavor as they wish.

珠海粤财假日酒店

珠海粤财假日酒店是一座30层的现代化酒店，坐落于珠海市繁华的金融商业中心地带，于2001年11月开业。距珠海拱北口岸广场15分钟车程，10分钟车程至九洲港码头（往返香港、深圳蛇口和珠海之间），45分钟车程至珠海机场。酒店提供免费穿梭巴士往返于九洲港码头与酒店之间。酒店的彩华轩中餐厅特聘请资深名厨献上正宗的粤菜、新派港式点心及特色名菜。卢奇欧意大利餐厅开放式厨房拥有意大利石窟比萨炉。班尼斯顿爱尔兰酒吧是珠海首家爱尔兰酒吧，可领略浓郁的爱尔兰风情。

FOSHAN

Hometown of Pottery and Gourmet Food

佛山

既是陶艺之乡，也是厨艺之乡

Foshan, a prefectural-level city in central south of China's southern province of Guangdong, was set up in 1983. The city administers five county-level districts-Chancheng District, Nanhai District, Shunde District, Gaode Destrict and Sanshui District-with a jurisdictional area of 3979.72 square kilometers.Rich in natural resources, Foshan is famous for five National Products of Geographical Identification: Shiwan ceramic dolls, gambiered Guangdong gauze, Heshui pueraria thomsonii benth, Sanshui black wax gourd and Leping cucumis melon, as well as other local specialties such as blindly public cake, marinated trotters, Jiujiang double distilled liquor, Daliang stewed milk beancurd, and Gaoming rice noodles etc.Foshan boasts quite a number of tourist attractions, including the "Foshan New Eight Sights" - Foshan Ancestral Temple, Nanfeng Kiln, Xiqiao Mountain, Nanhai Peach Garden, Qinghui Garden, ChenCun The World of Flowers, Gaoming soap screen Hill, and Sanshui Lotus World. Besides the New Eight Sights, another renown scenic spots include Dong Hua Li, Former Residence of Kang Youwei, Changlu Leisure Vocation Farm, Daqitou Ancient Village, to name just a few.

Foshan is a historic city with abundant inheritance of Lingnan culture. It's known as the home of Cantonese opera, martial arts, Guangdong gauze, traditional Chinese medicine patent prescription, folk arts, as well as the center of the southern foundry industry. It holds many cultural and religious festivities every year, like the Foshan Autumn Festival, the Tongji Bridge Parade, the Lettuce Gathering (lettuce in Chinese is a homonym for wealth generation).

Foshan is the cradle of folk arts in the Pearl River Delta region, and gives birth to many well-preserved folk arts and activities with Lingnan cultural features. A total of 13 Foshan folk arts are on the China National Non-Material Cultural Heritage List. They are Lion Dance, Cantonese Opera, Dragon Boat Rap, Foshan New Year Wood-Block Prints, Guangdong Paper-Cutting, Shiwan Pottery, Foshan Festoon Lion's Head, gambiered Guangdong gauze dying, Foshan Ancestral Temple Fair, Foshan Autumn Festival, Shifan Music (the ensemble of ten traditional Chinese percussion instruments), Human Dragon Dance, and Color Lantern. So far, eight Foshan locals are named Inheritor Representatives of China's National Non-material Cultural Heritage.

佛山市位于广东省中南部。1983年设立地级市，辖禅城区、南海区、顺德区、高明区、三水区，行政区域面积3 979.72平方公里。佛山自然资源丰富，主要名优土特产有石湾公仔、顺德香云纱、高明合水粉葛、三水黑皮冬瓜、乐平雪梨瓜等5个国家地理标志产品，以及盲公饼、酝扎猪蹄、九江双蒸酒、大良双皮奶、高明濑粉等；旅游景点有祖庙、南风古灶、西樵山、南国桃园、清晖园、陈村花卉世界、高明皂幕山、三水荷花世界等“佛山新八景”，还有岭南天地（东华里）、康有为故居、长鹿农庄、大旗头古村等知名景点。

佛山历史悠久，岭南文化浓厚。素有陶艺之乡、粤剧之乡、武术之乡、广纱中心、岭南成药之乡、南方铸造中心、民间艺术之乡等美誉，形成秋色、行通济、生菜会等独具特色的民俗。佛山是珠江三角洲民间艺术的摇篮，孕育并保留了大量体现岭南文化精髓的民间艺术和民俗事象。狮舞、粤剧、龙舟说唱、佛山木版年画、广东剪纸、石湾陶塑技艺、佛山狮头、香云纱染整技艺、祖庙庙会、佛山秋色、十番、人龙舞和佛山彩灯等13个项目入选国家非物质文化遗产名录，8人入选国家级非物质文化遗产代表性传承人。

Shunde, Hometown of Kongfu Masters and Master Chefs

Shunde is one of the five administrative divisions of Foshan City. It sits in the south of Guangdong Province, center of the Pearl River Delta Plain, south of the Guangzhou-Foshan-Zhaoqing Economic Cooperative Zone, and serves as one key hub linking the cities of Foshan and Guangzhou. It's always been known for its developed economic and commercial prosperity ever since ancient times. It once earned the reputation of "Guangdong Four Tigers" and ranked the first among China's Top 100 Districts for 2014.

As one major birthplace of Cantonese food, Shunde is known as the "Home of Chinese Gourmet" and "Cradle of Lingnan Cantonese Cuisine" for its long history, profound culture, diversified cooking materials and methods along with its sound food industrial base. It is also among China's most well-known places for its concentration of celebrity restaurants and celebrity chefs.

In 2004, Shunde was awarded "Home of Chinese Chefs" by the China Cuisine Association (CCA). In 2006, Daliang County under the District was named China's first Town of Catering. In 2010, Shunde was honored as the first "City of Catering in China" by the CCA, which in turn established and enhanced its position as the center of southern Chinese food culture.

Shunde dishes are made of fresh materials by the integration of perfect temperature control and various cooking methods for the best of color, fragrance, taste and shape. With tastes varying from sourness, sweetness, bitterness, hot, saltiness to freshness, Shunde cuisine forms its uniqueness of being delicate, tender, smooth, cool and fresh.

Sticking to the principle of eating with the seasons, Shunde cuisine pays attention to changing food tastes with seasons and is dedicated to being nutritionally scientific, health conducive and preserving, as well as beauty enhancing. Shunde dishes taste delicate in summer and intense in winter as chefs cook whatever available at the time so as to create the most natural and fresh tastes.

Shunde pasties are known for their sophisticated craftsmanship, delicate shapes and delicious crispness. A total of 11 snacks, including stewed mild beancurd, are among the "Famous Snacks in China". Meanwhile, the District is blessed with culinary talents. It has 23 celebrity chefs who are selected "Cooking Master in China" and 22 "China's Famous Cooks".

顺德——李小龙的故乡盛产厨师

顺德区是佛山市五个行政辖区之一。位于广东省南部、珠江三角洲平原中部、广佛肇经济圈南部，是佛山市与广州市联系的重要核心区域之一。自古经济发达，商业繁荣，曾有“广东四小虎”“2014中国百强区第一名”等美誉。

作为粤菜的主要发源地之一，顺德以其悠久的历史、厚重的文化底蕴、多样的烹调材料与方法以及坚实的产业基础而被誉为“中国美食之乡”“岭南粤菜之源”，是全国餐饮名店、名师最密集的地区之一。2004年顺德被中国烹饪协会授予“中国厨师之乡”称号，2006年顺德所辖大良镇获评国内首个“中华餐饮名镇”，2010年顺德成为中国烹饪协会颁发的中国首个“中国美食名城”，从而进一步确立和巩固了中国南方美食文化中心的地位。

顺德菜选料新鲜，讲求火候以及各种烹调技法的综合运用，在味道方面调味遍及“酸甜苦辣咸鲜”，菜式“香、色、味、形”均达到最佳效果，形成了“清、嫩、滑、爽、鲜”的特色。顺德菜讲究应时而食，重视口味的时令季节变化，讲求科学营养、健康有益和美容养生的效果。夏秋力求清淡，冬春则偏重浓郁，因地制宜，形成自然新鲜的美妙滋味。

顺德点心亦以制作精细、外形雅致、松脆可口见长，有双皮奶等11种名特小吃入选“中华名小吃”，成为顺德美食中的另一朵奇葩。餐饮人才更是群星璀璨，拥有23位“中国烹饪大师”和22位“中国烹饪名师”。

▸ Link: Recommended restaurants

Foshan Fortuna Hotel

Located in Lecong, junction of Shunde District and Chancheng District, Foshan Fortuna Hotel is adjacent to Lecong Cultural Park, National Highway 325, Sanle Road, and Foshan's first ringroad. Featuring a convenient transportation, it takes only 15 minutes to get to Foshan Railway Station and Foshan Bus Station, 45 minutes to the Shunde Port, and 60 minutes to Guangzhou Baiyun International Airport and China Import and Export Fair Complex.

Fortuna Hotel's interior design is a combination of the Chinese and Western arts, with exquisite handcrafts creating a magnificent and elegant atmosphere. All the dining halls are uniquely designed. Besides general accommodation, Fortuna Hotel also provides a string of five-star services such as free shuttle between the hotel and Foshan Ancestral Temple, 24-hour room service, whole-day banquet catering service, professional meeting and exhibition planning, specialized secretaries and business supporting services, all reflecting Fortuna Hotel's commitment to services of the highest level. Centering on the guests, Fortuna Hotel is always at their service.

佛山市财神酒店

佛山市财神酒店位于佛山顺德区与禅城区的交会处——乐从，毗邻乐从文化公园、325国道、三乐路、佛山一环，15分钟内便可到达佛山市火车站及汽车站，45分钟可到达顺德港，60分钟可到达广州白云国际机场及中国进出口商品交易会会场，交通四通八达。

财神酒店室内设计糅合中西艺术精髓，细致的手工，营造出宏伟而高贵的气氛，所有餐厅都有着其独特的设计风格。充分显现出个性化服务的精髓，除了为客人提供一般的酒店住宿，更提供一连串高质量的五星级服务，包括往返佛山祖庙与酒店的免费穿梭巴士、24小时的客房送餐服务、全天候的宴席到会服务、专业的会议及展销会策划、专业的秘书及商务配套服务等，都是财神酒店追求最高品质服务的真诚体现。

DONGGUAN

Hometown of Aquilaria Sinensis and a Food Paradise

东莞

莞香的故乡，美食的天堂

Located in south central Guangdong Province, Dongguan has been a prefecture-level city since 1988. Statistics show the city governs 28 towns and four sub districts. Geographically, Dongguan has 2,465 square kilometers of land area, 97 square kilometers of maritime area, and 115.95 kilometers of coastline. Altogether 1,668 kinds of wild vascular plants, categorized into 797 genuses and 204 families, can be spotted in the city. It is also a habitat to rear plants like Cyperus malaccensis, agarwood, Pinus fenzeliana, Lysidice brevicalyx, and Cephalotaxus fortune. Famous specialties include lychee, Dongguan sausage, Baisha dried salted duck, Mayong banana, Humen shrimp, Humen fish, Humen cherimoya, Wanjiang dried bean milk cream in tight rolls, and pomelo peel in malt sugar.

As a historic city in Guangdong Province, Dongguan enjoys a reputation of cradle of Chinese folk art, motherland of Chinese dragon dance, hometown of weight lifting, city of swimming, and birthplace of dragon boat. Well-known tourist attractions are the ruins of the Opium War-- the pits where Lin Zexu had opium destroyed, Weiyuan batteries, and Shajiao batteries, and patriotism education bases such as the Opium War Museum, and the Sea Battle Museum. Notable cultural attractions are Ying'enmen Gate Tower built in the Ming Dynasty, Jinaozhou Tower, Pomegranate Tower, Yuan Chonghuan's former residence, ancient Nanshe Village, Yanling crag inscription dating back to the Ming and Qing dynasties, Keyuan Garden--one of the four most famous gardens in Guangdong during the Qing Dynasty, Jiang Guangding's former residence, Guangdong Memorial Hall of East River Column. Spectacular landscapes are Buddha Mountain National Forest Park, Songshan Lake, Shuilian Mountain Forest Park, Botanic Garden, Chang'an Lotus Mountian, and Huying Country Park.

东莞市位于广东省中南部，1988年升格为地级市，辖28镇4街道。全市陆地面积2 465平方公里，海域面积97平方公里，海岸线长115.95公里。有野生维管束植物204科797属1 668种，珍稀植物有短叶茳芏（莞草）、土沉香（莞香、白木香）、华南五针松（广东松）、短萼仪花（麻忆木）、三尖杉等。土特产有荔枝、东莞腊肠、白沙油鸭、麻涌香蕉、虎门麻虾、虎门水鱼、虎门番荔枝、万江腐竹、麦芽糖柚皮等。

东莞市是广东省历史文化名城，中国曲艺之乡、中国民间艺术之乡、龙狮之乡、举重之乡、游泳之乡、龙舟之乡。主要旅游景区（点）有林则徐销烟池、威远炮台、沙角炮台等鸦片战争遗址和鸦片战争博物馆、海战博物馆等爱国主义教育基地，明代的迎恩门楼、金鳌洲塔、榴花塔、袁崇焕故居、南社古村落、明清时期的燕岭摩崖石刻、清代广东四大名园之一的可园和蒋光鼐故居、东江纵队抗日纪念馆等人文景点，有广东观音山国家森林公园、松山湖景区、水濂山森林公园、植物园、长安莲花山以及虎英郊野公园等自然景观。

▸ Link:
Recommended
restaurants

Hyatt Regency Dongguan

Hyatt Regency Dongguan is Guangdong's first five-star hotel built by the global hospitality company Hyatt International Corporation. It is located in the business district of the Songshan Lake High-tech Industrial Development Zone. Adjacent to the Songshan Lake, this uniquely designed hotel offers a breathtaking lakefront view, and is a popular choice for visitors in the Pearl River Delta Area.

The hotel offers a spectacular view of the Songshan Lake and lychee garden from its multi-functional banquet hall which covers 861 square meters. With seven multi-functional meeting rooms, a 2,500 -square meter multi-functional convention center, and a banquet hall capable of throwing a cocktail party attended by up to 1,000 people, all equipped with sophisticated conference facilities, Hyatt Regency Dongguan can satisfy the demands for various meetings and events.

东莞松山湖凯悦酒店

东莞松山湖凯悦酒店是著名的凯悦国际酒店集团在广东省的首家五星级酒店，位于松山湖高新技术产业开发区行政商务中心。酒店倚风景秀美的松山湖而建，四周碧水环绕，结合国际顶级大师别具匠心的设计，是珠三角地区难得的商务、休闲度假胜地。

占地861平方米的多功能宴会厅，可俯瞰松山湖极致美景和荔枝园林景观，另有7间多功能会议厅及2 500平方米的多功能会议展览中心，附设一间能承接1 000人鸡尾酒会的宴会厅，均配备了尖端会议专用设施，灵活配合各类会议及活动的需求。

Cymbidium House

Cymbidium House is located on the fourth floor of the eighth part of the First International Plaza, No.200 of the Hongfu Road of Dongguan's Nancheng District. It is adjacent to the landmark buildings such as the office building of Dongguan Municipal Government, Dongguan International Exhibition Center, and Dongguan Yulan Theatre. With a pleasing environment, delicate desserts, and tasty sea food, Cymbidium House is a popular place to try the Cantonese cuisine.

芷兰轩正宗粤菜

芷兰轩正宗粤菜位于东莞市南城区鸿福路200号第一国际八区4楼（会展中心正门对面），靠近东莞市政府、会展中心、白玉兰大剧院等地标性建筑，是粤式饮茶的热门去处，环境温馨，点心出品精致，海鲜也可圈可点，喝茶要等位也就不足为奇了。

ZHAOQING

Gourmet Food as a Tourism Product

肇庆

美食是重要的"旅游产品"

With a history of more than 2,200 years, Zhaoqing was listed in the third batch of national historic and cultural cities. It is the political, economical and cultural center of Xijiang River region and a place of strategic importance. As the cradle of Lingnan Culture and Cantonese Culture, Zhaoqing is the earliest crossroad between Western civilization and traditional Chinese culture. Its Plum Temple, Zhaoqing ancient city wall, Seven Star Crag Inscriptions, and Longmu Ancestral Temple are dubbed National Key Cultural Relics Protection Units. Zhaoqing also has a total of 240 provincial, municipal (county) cultural relics protection units. Fengkai People, Guangxin Culture, Duan Inkstand Culture, Duanzhou Culture, Underground Forest, Yandu Forest, and Guangzhu Culture are among the list of Guangdong's first batch of Pearl River Cultural Constellation, and Duan Inkstand is one of the ten calling cards for Lingnan Culture.

肇庆是第三批国家历史文化名城，有文字记载的历史2 200多年；是岭南土著文化和广府文化发祥地，也是中原文化与岭南文化、西方文明与中国传统文明最早的交会处；曾是西江流域政治中心和军事重镇。梅庵、德庆学宫、肇庆古城墙、七星岩摩崖石刻、悦城龙母祖庙是国家级重点文物保护单位。省、市（县）级文物保护单位240处。“封开人”“广信文化”“端砚文化”“端州文化”“地下森林之都”“燕都文化”和“广竹文化”入选广东省首批“珠江文化星座”，端砚入选“岭南文化十大名片”。

Favorable natural conditions are nourishing unique creatures like Wendao carp, Maixi carp, mountain goat, and Xinghua broiler. Embodying Lingnan customs, Zhaoqing foods are based on high quality materials and are mostly famous for stewed dishes. A trip generally involves eating, housing, travelling, sightseeing, shopping and entertainment. That's to say, delicacy is one of the most important tourism products.

Zhaoqing's ten specialties are euryale ferox salisb, Zhaoqing rice dumpling, Hylocereus undatus, Deqing tribute orange, Sihui orange, Wenqing carp, Guangning bamboo core tea, Fengkai millet, Huaiji 60-day day-lily buds, and Maixi grass carp. It also has eight famous artwares, namely Duan inkstand, Sihui jadeware, Gaoyao wood carving, Seven Star red bean decoration, Choerospondias axillaris, Guangning emerald, Guangning bamboo carving, and Deqing master token.

肇庆，山湖江汇聚，优越的自然条件孕育出丰富的物产——文刀鲤、麦溪鲤、石山羊、杏花鸡，数不胜数。优质的食材资源让肇庆菜充满岭南风情，个性彰显，它注重质和味，清则油盐白水，浓则五味交融，尤以焖煲类出品最具特色。食住行游购娱，美食是一项重要的“旅游产品”。

肇庆还评出了十佳旅游手信：肇实、肇庆裹蒸粽、七星剑花、德庆贡柑、四会砂糖橘、鼎湖文庆鲤（鱼干）、广宁竹芯茶、封开油栗、怀集六十日黄菜、高要麦溪鲩（鱼干）。八大旅游工艺品：端砚、四会玉器、高要木雕、七星岩红豆饰品、五眼佛果、广宁绿玉、广宁竹雕、德庆状元牌。

EASTERN GUANGDONG

粤东

Yuedong region refers to the eastern part of Guangdong Province, including four prefecture-level cities (Shanwei City, Jieyang City, Chaozhou City and Shantou City) and two sub-prefecture-level cities (Shenzhen-Shantou Special Cooperation Area and Jinning City). The region has an area of 15,516 square kilometers. According to the data from the sixth population census in China, the region has a population of 17,172,100.

In 2014, the GDP of the four cities of Yuedong Region totaled 506.365 billion yuan, a year-on-year increase of 9.52 percent. Yuedong Region has become an important growth pole of Guangdong's economic development.

The local diets in Yuedong Region are mainly Guangdong cuisine, including Chaozhou and Shantou dishes.

粤东地区，是广东省东部地区的简称，包括汕尾市、揭阳市、潮州市、汕头市四座地级市和深汕特别合作区、普宁市两个副地级市。总面积15 516平方公里，全国第六次人口普查常住人口为1 717.21万。

2014年，粤东四市GDP共计5 063.65亿元，同比增长9.52%，粤东已成为广东经济发展的重要增长极。

粤东地区的本地饮食以潮汕菜为代表。

SHANTOU

Hometown of Overseas Chinese and a City with Exquisite Food

汕头

开放的侨乡，考究的菜肴

Shantou city is located in the eastern part of Guangdong Province. With a total administrative region of 2064.4 square kilometers, it comprises of Jinping, Longhu, Chenghai, Haojiang, Chaoyang, Chaonan Districts and Nan'ao County. The Shantou Bonded Area and Shantou High and New Technology Industrial Development Zone have been established. It is widely known that many overseas Chinese are from Shantou. As it lies in the marine outfall of Han, Rong and Lian Rivers, Shantou is one of China's pioneering port cities in modern times and also a famous trading post. Dubbed "A Stronghold in Eastern China and Eastern Gateway to Guangdong," Shantou has been a crucial transportation hub, import and export port and logistics center in eastern Guandong, southern Jiangxi and southwestern Fujian Provinces for a long time.

Shantou is renowned for the abundance of Lingnan fruits. Among them, the orange of Chaozhou is dubbed the "queen of oranges." The local fruits like apple, olive, red bayberry, emblic leaf flower, and the Nan'ao pomegranate are all of the high quality. It is also known as "the kingdom of vegetables" where the vegetables in season are of the best quality. So in Shantou one could find the superior leaf mustard, Chinese cabbage, mustard blue, and cauliflower. Because of this, the vegetable seeds of Shantou are famous both in China and in overseas countries.

The coastline length of Shantou, including that of the islands, totals 289.1 kilometers. It consists of forty islands in all with Nan'ao Island as the largest. So far as we know, there are 471 kinds of sea fishes, 17 kinds of shrimps and crabs, 30 kinds of shellfishes, 20 kinds of algae. There is a variety of marine life both in the sea and in the freshwater, such as eel, rudd, pomfret, sierra, prawn, crabs, portunid, squid, oyster and nori and many others. Chaozhou cuisine is well-known around the world for delicate materials, exquite making process and distinctive flavor. The local food is represented by baked oysters and three kinds of meat balls: stewed beef meatballs, pork meatballs, fish balls. Varieties of succades and preserved vegetables are also widely known as Shantou produce. The handicrafts are of high reputation both at home and abroad, such as Chaoshan drawn work, the beaded ornaments, wood carving, painted porcelain, gold and silver ornaments.

汕头市位于广东省东部，辖金平、龙湖、澄海、濠江、潮阳、潮南6个区和南澳县，另设立汕头保税区和汕头高新区，行政区域面积2 064.4平方公里。是全国著名的侨乡。汕头地处韩江、榕江、练江出海口，是近代中国最早对外开放的港口城市和著名商品集散地，历来是粤东、赣南、闽西南一带的重要交通枢纽、进出口岸和商品集散地，素有“华南之要冲，粤东之门户”的美称。

汕头盛产各种岭南佳果，潮州柑被誉为“柑橘皇后”，苹果、橄榄、杨梅、油柑、澳榴等是当地名优水果。又是“蔬菜王国”，四季均有优质当令蔬菜应市，包心芥菜、早皇白菜、芥蓝花、椰菜花是当地特优品种，良种菜籽蜚声海内外。

全市海岸线和岛岸线长289.1公里，有大小岛屿40个，最大的是南澳岛。已知近海鱼类有471种，虾蟹类17种，贝类30多种，藻类近20种。海、淡水产品丰富，如鳗鱼、石斑鱼、鲳鱼、马鲛鱼、对虾、膏蟹、梭子蟹、鱿鱼、牡蛎、紫菜等。潮州菜以其用料考究、制作精巧、美味独特而风靡海内外。以牛肉丸、猪肉丸、鱼丸“三丸”和蚝烙等为代表的风味食品，远近闻名。潮汕抽纱、金漆木雕、彩瓷、珠绣、金银饰品等手工艺品享誉海内外。

▸ Link: Recommended restaurants

Haiji Beef

Just like stewed beef meatballs, the beef hot pot is another local cuisine. Haiji Beef is the most characteristic restaurant in cooking such cuisine. (Address: No. D 14 – D 16, Block 6, Huanggang Road, Shantou City; close to the Baheli Lane). Some visitors even have made such comment that it would not be a worthwhile trip to Shantou without tasting Haiji beef. On the menu the prices of different parts of the beef have been written down, such as Diaolong (sirloin or fillet), Shaobing (the tenderloin), Longxia Xu (the meat in the cracks in the basin). The "Boren" or "Niufeng" is highly recommended, with the chuck tasting excellently in beautiful "marble patterns".

海记牛肉

海记牛肉位于汕头市黄岗路坪西6座014—016，近八合里巷。跟牛肉丸一样，牛肉火锅是汕头的美食名片，而海记牛肉就是这张名片上名字最响亮的，甚至有饕餮之客说：没吃过海记牛肉就等于没到过汕头。菜单上写着牛肉的不同部位和价格，比如吊龙、匙柄、龙虾须等。尤其值得推荐的是“脖仁”或“牛峰”，是一种有好看的“大理石状花纹”的牛颈肉，口感极佳。

CHAOZHOU
Hometown of Kongfu Tea

潮州
工夫茶的故乡

Chaozhou is located in the eastern part of Guangdong Province. It was established as a prefecture-level city in 1991, it comprised Xiangqiao District, Raoping County, Chao'an County, and a Fengxi District Committee was established separately. The total area of the city is 3679 square kilometers, including the land area of 3146 square kilometers, the sea area of 533 square kilometers. Rich in historical and cultural heritages, it is dubbed "the Seaside Zou Lu" (Zou Lu in Chinese literally means a place with prosperous culture), and "the marvelous city in the areas of Guangdong and Guangxi."

The local specialties of Chaozhou include the Chaozhou cuisine, the Chaozhou snacks, the Phoenix Tea, Lingtou Tea, Chaozhou tangerines, and various kinds of succades. The traditional artistic work includes ceramics, embroideries, drawnworks, wood carvings and straw paintings. Chaozhou is a key national arts and crafts city, the capital of arts and craft across China, a national daily ceramic industry base, a ceramic export base and an international procurement base of clothes in China. All through the city, there are 1345 historical relic sites, among which there are 8 important historical monuments under special preservation: the Guangji Bridge (Xiangzi Bridge), Xu Fuma Palace, Kaiyuan Temple, JilueHuanggong Temple, the Relics of Song Dynasty kiln in Bijia Mountain, Hanwengong Temple, DaoyunTulou and the Congxigong Temple. The Yixingjia Lane on Chaozhou Taiping road is a famous Chinese historical and cultural boulevard. Besides, Chaozhou boasts beautiful natural sceneries. The main scenic spots include the Green Island Hills, Dongshan Lake Thermal Spring Resort, Danfu Cultural Relics Collecting Institutes, Scenic areas of the West Lake, Guangdong Hongshan Forrest Park and Guangdong Zilian Mountain Forrest Park.

潮州市位于广东省东部。1991年设立地级市，辖湘桥区、饶平县、潮安县，另设枫溪区管委会。全市总面积3 679平方公里，其中陆域3 146平方公里，海域533平方公里。历史文化底蕴深厚，素有“海滨邹鲁”“岭海名邦”之称。

潮州的地方特产主要有潮州菜、潮州小食、凤凰茶、岭头茶、潮州柑和各式凉果等。传统工艺品主要有陶瓷、潮绣、抽纱、木雕、麦秆画等。潮州是国家重点工艺美术城市、中国工艺美术之都、国家日用陶瓷特色产业基地、中国陶瓷出口基地和中国服装跨国采购基地。全市有不可移动文物1 345处，其中有广济桥（湘子桥）、许驸马府、开元寺、己略黄公祠、笔架山宋窑遗址、韩文公祠、道韵楼、从熙公祠等8处全国重点文物保护单位。潮州太平街义兴甲巷是中国历史文化名街。潮州自然风光秀丽，主要景区有绿岛山庄、东山湖温泉度假村、淡浮院、西湖风景名胜区、广东红山森林公园和广东紫莲山森林公园等。

▸ Link:
Recommended restaurants

The Century-old Hurongquan

Huronquan is the century-old dessert shop (Address: No. 136 – 144 of Taiping Road). Yamunian is the specialty of the shop. It is a dessert mainly made by glutinous rice. The sweet dumplings with glutinous rice flour are made as the wrapper, with mung bean paste, taro, sesame and red bean paste as the fillings. Then the dumplings are boiled with tremella, barley, red dates, sweet potato stems and lilies in the syrups. When finished, the dumplings are ladled out with syrups. So the taste is sweet but not cloying. One bowl is sold with 8 yuan. There are also fresh and tasty wontons sold in the morning.

百年老店胡荣泉

潮州的百年老店胡荣泉位于太平路136—144号，鸭母稔最有名。鸭母捻是一种甜品，糯米粉做成的汤圆分别包上绿豆沙、芋泥、芝麻、红豆沙四种馅儿，放入加了银耳、薏米、红枣、红薯干、百合熬成的糖水里煮熟，就着糖水盛出来，甜而不腻，8元一份。早上也可以过来吃馄饨，入口即是鲜味。

WESTERN GUANGDONG

粤西

The western Guangdong area comprises three prefecture-level cities, namely Zhanjiang, Maoming and Yangjiang. However, the area will only include the first three cities in terms of economic development.

Most areas of western Guangdong locate on mountains and hills, but Zhanjiang and Maoming lie on tableland area. The region spans 31,700 square kilometers, accounting for 17.7 percent of the total area of Guangdong Province. The number of permanent residents living in western Guangdong in 2014 was nearly 15.66 million.

Western Guangdong's GDP achieved 577.6 billion yuan (about US$93.1 billion) in 2014, a 9.82 percent year-on-year increase. The area has become another driving force for Guangdong's economic growth.

粤西（粤西地区）是广东省西部地区的简称，按经济发展角度划分，包括湛江、茂名、阳江三市。

地形以山地丘陵为主，湛江、茂名属台地。区域总面积3.17万平方公里，占全省的17.7%；2014年常住人口1 565.92万。

粤西地理位置优越，处于珠三角与北部湾、大西南联结点以及粤港澳合作次前沿。2014年GDP为5 776.3亿元，同比增长9.82%，已成为广东经济的另一增长极。

ZHANJIANG

China's No.1 Seafood City

湛江

中国第一海鲜美食之都

Being surrounded by the sea on three sides, Zhanjiang City lies on Leizhou Peninsula in the southwestern end of Guangdong Province. It is seen as the thoroughfare of Guangdong, Guangxi and Hainan. As one of the starting point of the Maritime Silk Road, Zhanjiang is one of China's first coastal cities to implement the policy of opening up to the outside world and has been crowned with titles such as China's Excellent Tourism City, National Garden City, and Homeland of China's Seafood Delicacies. Zhanjiang has a total of 1,556 km of coastline and has more than 140 islands and bays. It is a typical coastal city with the longest coastline and most islands in China. Excellent tourism resorts in the city has perfectly displayed its tropical and subtropical charm, such as the "Five Islands and One Bay" seaside resort ("Five islands" refers to Techeng Island, Nansan Island, Donghai Island, Naozhou Island and Nanping Island. "One Bay" refers to Zhanjiang Bay), China's longest Longhaitian Beach, Xuwen National Coral Reef Nature Reserve, Zhanjiang Mangrove National Nature Reserve, Leizhou Rare Marine Creature National Nature Reserve, Naozhou Island (China's largest volcanic island), Dahan Sandun tourist attraction (The departure of the Maritime Silk Road in Han Dynasty), and Tianchengtai Resort (a perfect place to watch sunset).

湛江市位于中国大陆最南端，地处我国三大半岛之一的雷州半岛，居粤桂琼三省区中枢，三面环海，拥有中国唯一的热带半岛，是海上丝绸之路始发地之一、国家首批对外开放沿海城市、中国优秀旅游城市、国家园林城市、中国海鲜美食之都。湛江拥有长达1 556公里的海岸线、140多个岛屿，海岛、海湾星罗棋布，是广东省海岸线最长、海岛最多、海味最浓的地级市。这里有“五岛一湾”特色滨海旅游区、中国第一长滩——龙海天沙滩、徐闻国家级珊瑚礁自然保护区、红树林国家级自然保护区、雷州珍稀海洋生物国家级自然保护区、中国最大的火山岛硇洲岛、汉代海上丝绸之路始发港——大汉三墩、中国最美的落日观赏地——雷州天成台，尽显依海而生的热带、亚热带城市魅力。

ʾeople in Zhanjiang City make their living from the sea. Everything hey do in the city is connected with the sea, such as bathing, ʾishing, eating, sightseeing and shopping. Consequently, the ɛeafood culture has a long history in Zhajiang. "Silent on teeth, ɛoft on tongue"--Zhanjiang's seafood delicacies have been highly ɔraised as such by celebrities since ancient times. They are truly a feast for the palate due to rich varieties, fresh ingredients and ophisticated cooking techniques.

Zhanjiang is also famous for its fruits. There are wide range of ʾruit production base for growing pineapples, mangoes, Hongjiang oranges, red carambolas, longans, lychees and green jujubes. The city can provide fresh fruits in any season. Moreover, as the south-o-north vegetable transportation base, Zhanjing's hot pepper, cucumber and eggplant have been sold across the country.

湛江以海为生，饮食文化源远流长。“入齿脆无声，坐令吾舌柔。”“雪色一寸余，脱甲琼浆流。”湛江的海鲜在古代就已得名人志士的盛赞。“洗海浴、钓海鱼、吃海鲜、观海景、买海货”，以海鲜美食为龙头的湛江美食，品种丰富，选料鲜活，做法讲究而富于变化，堪称味蕾上的饕餮盛宴。

湛江是水果之乡，有连片面积广阔的菠萝、芒果等种植基地，有国宴珍品红江橙、“岭南佳果”红杨桃、龙眼、荔枝、黄皮、青枣等，优质水果数不胜数，一年四季水果飘香。这里是南菜北运基地，辣椒、青瓜、紫茄等瓜菜畅销全国各地。

MAOMING

A Place Where Food Culture is more Profound than Guangzhou

茂名

饮食之风，甚于广州

Maoming City, famous for its splendid landscapes and abundant tourism resources, is located at the southwestern part of Guangdong Province. On the one hand, there are a number of beautiful natural landscapes in the city, such as Dawuling Nature Reserve, Tianma Mountain, Xijiang Hot Spring Resort, the First Beach Tourist Resort, the Hutou Mountain Park, the Longtou Mountain Forest Park, Fushan Mountain and Magang Hot Spring. The extraordinary scenery of The First Beach Tourist Resort, also known as the "Oriental Hawaii" and the "Beidaihe River in South China", has been painted as the background of the "Green Great Wall", a well-known landscape painting hanging in the Great Hall of the People. The waters around Fangji Island are the world's second cleanest, with underwater visibility ranging from 8 to 10 meters. Therefore, Fangji Island is also honored as the "Rare Underwater Park". On the other hand, many historical cultural relics can be found in Maoming City, such as Xian Imperial Ancestral Temple in Gaozhou City, Guanyin Temple, Baoguang Pagoda, the Ancient Dianbai County Relics, the Deluge Palace Site in Yixin City, and the Drum Tower in Huazhou County. In addition, extensive production bases for lychees and longans have become ecological agriculture tourist attractions in Maoming.

茂名市位于广东省西南部，风光秀丽，旅游资源丰富。自然景观有大雾岭省级自然保护区（主峰大田顶海拔1 704米，为广东省第二高峰）、天马山、西江温泉、第一滩、虎头山、龙头山、浮山岭、麻岗热水温泉等。水东第一滩蓝水白沙气势恢宏，素有“东方夏威夷”“南方北戴河”之称，北京人民大会堂悬挂的《绿色长城》就是以此为背景，由著名画家关山月创作；放鸡岛周围海域水下能见度为8米，居世界第二位，被誉为罕见的“海底公园”。人文旅游景点有高州冼太庙、观音寺、宝光塔、古电白郡遗址、信宜大洪国王宫、化州古楼等。此外，区域化连片的荔枝、龙眼生产基地已成为生态农业旅游景点。

Behind mountains and facing the sea in front, Maoming has a long history and profound food culture. The most quintessential food in the city is the Cantonese cuisine, but Maoming also has its distinctive cooking skills. In terms of different geographical conditions, cultures and customs, Maoming's delicacies, with characteristics of light flavors, fall into three categories in: Dianbai cuisine, Huazhou cuisine, Xinyi cuisine and Gaozhou cuisine.

People's diet in Baidian, one of Maoming's districts by the sea, is seafood. All of their dishes have light flavor and taste mellow, soft and fresh. Xinyi and Gaozhou are two cities lying on mountainous and hilly areas, so most food materials are derived from mountains. The authentic flavors of food and heat control are important during cooking procedures. "He", a kind of pastry made from rice and wheat with various fillings, is the most iconic traditional food in Gaozhou. People usually make various shapes of "He" in different flavors during "solar terms" and festivals. In Huazhou City, the most popular dishes are boiled chicken slices, steamed chicken and beef brisket, which keeps the authentic flavors of chicken and beef without greasy.

茂名背山面海，历史悠久，饮食文化丰富，以粤菜风味为主，又有自身独特的烹制之法。茂名饮食以清淡鲜活为主，而由于各地地理环境、人文风俗不同，又分为电白菜、化州菜、信宜菜和高州菜。

茂名电白区近海，食材以海鲜为主，菜品以清、香、嫩、鲜为主。信宜和高州多山地丘陵，饮食具有鲜明的山野气息和乡村特色，以农家菜为主，突出主料，原汁原味，讲究酥软香浓，注重火候。高州传统美食中，粄最具特色。在各个时令节气、喜庆时日，高州人都要做粄，而且每个节庆都做不同的粄，如寿桃粄、簸箕炊，各有风味。化州则以白斩鸡、隔水蒸鸡和牛腩出名，追求原汁原味，味道香重而不腻，更加接近广州菜。

YANGJIANG

Visit the Ancient Port of the Maritime Silk Road and Taste Fresh Seafood

阳江

游海丝古港，品天然海味

Yangjiang City locates in the southwestern part of Guangdong Province. After several subsequent shifts of jurisdiction, it became Yangjiang prefecture-level City in 1988. It comprises Jiangcheng District, Yangdong County and Yangxi County, Hailing Island Economic Development Pilot Zone and Yangjiang High-Tech Industrial Development Zone. It has an administrative area of more than 7,955 square kilometers (urban area of 433.7 square kilometers). Yangjiang has been honored as the Fishing Capital of South China Sea. Zhapo and Dongping, two counties in Yangjiang, are national level fishing ports. Some 105 types of seawater fish have been found to have economic values. As China's No. 1 sea products producer, Yangjiang City is rich in various species of seafood, such as squid, jelly fish, flower crab, prawn, mackerel and grouper.

Yangjiang has many kinds of special local products, including knives and scissors, kites, salt black beans, painted products, the malachite, amomum villosum fruits, yellow mane gooses, jade pouch lychees, Manshui oranges, the alpinia oxyphylla, the fried rice biscuit, Zhuchanglu (Rice rolls with fillings) and the salted fish. Well-known tourist attractions include Guangdong Maritime Silk Road Museum, Zhapo Dajiao Bay Resort in Hailing Island, Dongping scenic area, Shapa Bay Haitian Resort, Lingxiaoyan scenic area in Yangchun City, Yuxi Three Holes scenic area, Chundu Hot Spring Resort, Yangjinag Hot Spring Resort, the Salt Mine Resort in Yangxi County, the Donghu tourist area, the Shoujing Resort, the Shibazi Tourism Industry Pilot, the Jingye Temple in Dadong Mountain, the Shijue Temple, the Beishan Park, the Yuangyang Lake Park and the "Treasure of the Country" Museum.

Dubbed as "Homeland of Culture", Yangjiang City has been thriving on folk culture and art activities since ancient times. Famous cultural activities include calligraphy, couplet, Xianshui Song (Xianshui literally means "saltwater". The "Xianshui Song" is a kind of folk song that has been popular in the Pearl River Delta.) and folk song competition. A large number of outstanding celebrities came from Yangjiang, such as Madame Xian, a political leader of the Yue ethnic group in Southern Dynasties; Ruan Tuizhi, a well-known calligrapher and poet honored as one of the "Eight Outstanding Artists" in Lingnan Area (Lingnan Area comprises Guangdong Province, Guangxi Zhuang Autonomous Region and Hainan Province) and Guan Shanyue, an outstanding contemporary painter who has made great contribution to the development of traditional Chinese paintings. Yangjiang has built the Maritime Silk Road Museum on Hailing Island, where the shipwreck of "Nanhai No. 1", a wooden vessel which sank off Hailing Island at the end of the 10th Century, has been found. Thousands of hundreds of tourists have been attracted to Hailing Island in bid to get a view of the ancient ship and enjoy delicious seafood delicacies.

阳江市位于广东省西南部。1988年设立地级市，辖江城区、阳东县、阳西县，代管阳春市，另设海陵岛经济开发试验区、阳江高新技术产业开发区，行政区域面积7 955.3平方公里（其中市区面积433.7平方公里）。阳江是中国南海渔都，闸坡和东平是国家级中心渔港。已发现有经济价值的海水鱼类品种105种。海水产品产量稳居全省首位，盛产鱿鱼、鱼翅、海蜇、花蟹、对虾、马鲛鱼、石斑鱼等。

阳江特产主要有阳江刀剪、阳江风筝、阳江豆豉、阳江漆器、孔雀石、春砂仁、阳江黄鬃鹅、双肩玉荷包荔枝、马水橘、益智、粉酥、猪肠碌、咸鱼等。主要旅游景点有广东海上丝绸之路博物馆、海陵岛闸坡大角湾旅游区、东平景区、沙扒海天度假村、阳春市凌霄岩风景区、阳春市玉溪三洞风景区、阳春市春都温泉、阳江温泉度假村、阳西咸水矿温泉度假村、阳东东湖旅游区、阳江涛景度假村、十八子旅游工业示范点、阳西大垌山净业寺、石觉寺、北山公园、鸳鸯湖公园、金山植物公园、国之瑰宝工艺馆等。

阳江是"文化之乡"。群众性的文化、艺术活动源远流长。书法、对联、咸水歌、山歌擂台闻名遐迩。历史上有南朝越族首领冼夫人，近代有"岭南八子"之一的南国诗人、著名书法家阮退之，现代有著名国画大师关山月等文化名人。现在发现"南海Ⅰ号"沉船的海陵岛还建立了广东海上丝绸之路博物馆，吸引着游客到此赏沉船珍奇，品天然海味。

▸ Special Recommendation

Experience Hailing Island–Nimble•Happy Golden Island

The Nimble•Happy Golden Island, lying on Hailing Island which is known as one of China's "Top 10 Beautiful Islands", is a precious and marvelous tourism destination by the sea.

The ecological beach extends about five kilometers, providing you an excellent opportunity to experience the charm of nature. The Nimble•Happy Golden Island scenic area has been built on the most amazing part of the Hailling Island. It has retained the original ecological environment of the beach. The silver sand will sparkle in the sun as blue waves wash the seashore.

You can enjoy Asia's distinctive Crystal Lake and walk along the golden beach in the Golden Island. The 1-km long Crystal Lake, named for its crystal chain-like aerial view, has a total area of 80,000 square kilometers. There is a 20,000-square kilometer infinity edge pool in the lake. Soaking in the swimming pool over the beach, you may feel as if you were stay in the sea.

Don't miss the transparent canoe! You can relax through a variety of fashionable and exciting entertainment from around the world, such as traveling in transparent canoes, water skiing, operation sailing, windsurfing, riding motorboats, surfing, riding hoverboard, flying fish trip, playing beach soccer and volleyball.

The Nimble•Happy Golden Island apartment, adjoining the tourist area, has over one thousand harbor view apartments with diversified decorating styles, including that of the Southeast Asia, the Mediterranean, the North Europe and the New Chinese style. All the apartments are being managed by world's top-notch teams in bid to build China's best holiday apartments with the most beautiful harbor views.

海陵岛顶级体验——敏捷·欢乐黄金岛

敏捷·欢乐黄金岛景区位于“中国十大美丽宝岛”之一的阳江海陵岛，成就中国滨海线上最壮观的“天造一半，人造一半”的瑰宝级旅游景点。

在这里，坐拥五公里生态海滩，贴近大自然，偶遇天然贝类。敏捷·欢乐黄金岛景区的海岸线是沿海陵岛十里银滩上最美的五公里生态海滩而建，保留了沙滩的原生态环境，沙滩细沙为原有沙石资源，幼细硅白，经碧蓝清澈的海水冲刷，在阳光下泛着粼粼波光。

在这里，饱览亚洲独有一公里透明水晶湖，漫步铂金沙带。敏捷·欢乐黄金岛水晶湖总面积达8万平方米，长约1公里，从天空俯瞰，仿若一条全透明的水晶链。水晶湖内同时还拥有2万平方米的无边际泳池，此处水面高于沙滩，有种与海平起平坐的神奇感受。

在这里，必玩项目为全岛独有的透明独木舟。景区特意从世界各地搜罗时下最新、最嗨的游玩娱乐项目：全岛独有透明独木舟、休闲滑水板、水上OP帆船、GP风帆。除此外，景区还拥有海上摩托艇、海上冲浪、海上飞板、飞鱼船、沙滩足球、沙滩排球等多个水上娱乐项目。

敏捷黄金岛度假公寓紧靠景区，拥超过千套的全海景公寓，并拥有“东南亚”“地中海”“北欧”“新中式”等多种不同主题风格房间。由国际顶级的公寓运营团队经营，目标是打造中国最美的全海景度假公寓。

Tourism hotline: 0622-8308889

Address: The Nimble Boulevard, Hailing Island, Yangjiang City, Guangdong Province.

Driving directions in Guangzhou: Guangzhou Outer Ring Expressway—Guangfo (Guangzhou-Foshan) Expressway (to Yangjiang)—Luoyang (Luoding-Yangjiang) Expressway (to Yangjiang Port)—Zhapo/High-Tech Zone Exit—Hailing Island (Turn left after entering the island)

旅游热线：0622-8308889

地　　址：广东阳江海陵岛敏捷大道

广州自驾游路线：环城高速—广佛高速（往阳江）—罗阳高速（往阳江港）—闸坡/高新区出口—海陵岛（入岛转左）

NORTHERN GUANGDONG

粤北

The northern Guangdong area comprises Shaoguan City, Qingyuan City, Meizhou City, Heyuan City and Yunfu City in terms of economic development. It spans 77,100 square kilometers, accounting for 42.9 percent of the total area of Guangdong Province. The number of permanent residents living in northern Guangdong was nearly 16.46 million in 2014.

The northern Guangdong area has a long history. It is an important hub for migration of the Guangfu clan and a key place for the formation and development of the Hakka clan. The Hakka cuisine is the most quintessential of food in northern Guangdong, but people in the city of Yunfu prefers the Cantonese cuisine.

粤北，即广东省北部地区，按经济发展角度划分，包括韶关、清远、梅州、河源和云浮五市。全区面积7.71万平方公里，占全省面积的42.9%。2014年常住人口1 645.68万。

粤北地区有悠久的历史，是广府民系南迁的重要中转站。同时，粤北也是客家民系形成与发展的重地。饮食以客家菜为代表。云浮的饮食则以广府菜为主。

SHAOGUAN

Hakka Flavor and Yao food

韶关

客家风味 + 瑶寨美食

Shaoguan City locates on the upper Beijiang River, in the northern part of Guangdong Province. Bordering Hunan and Jiangxi Province, the city is seen as the thoroughfare of the three provinces. It is the political, economic, transportation and cultural center for the northern Guangdong Province. It is also a commodity distribution hub for its neighboring provinces and regions. Shaoguan is rich in tourism resources. It was listed as one of Excellent Tourism Cities of China. The city has 17 world-class or state-level scenic spots and over 100 provincial-level and other scenic spots. You can find almost all elements of tourism here. Excellent ecological resources and rich cultural tradition have enriched Shaoguan tourism. The famous tourist attractions include cultural landscape such as Maba hominid site, Nanhua Temple, Zhuji Lane, Mantang Hakka Manor House, and The Ancient Meiguan Pass Road, as well as World Natural Heritage Danxia Mountain, Guangdong Nanling National Forest Park and Guangdong Grand Canyon.

韶关市地处广东省北部，北江上游，与湖南省、江西省交界，素有“三省通衢”之称，是粤北地区的政治、经济、交通、文化中心，粤湘赣交界地区商品集散中心。韶关旅游资源丰富，是中国优秀旅游城市，有世界级、国家级景区景点17处，省级及以下景区景点100多处，几乎囊括了旅游的所有元素。优越的生态资源和丰厚的人文传统，赋予了韶关旅游丰富的内涵。著名的旅游景点有马坝人遗址、南华禅寺、珠玑古巷、满堂大围、梅关古道等人文景观，以及世界自然遗产丹霞山、南岭森林公园、广东大峡谷等自然景观。

Shaoguan has a history of over 2,000 years. It is a popular Hakka settlement. The local people are simple and sincere. The unique folk custom, together with rich food ingredients, has cultivated a lot of unique local delicacies. Hakka cuisine can be divided into two categories: Hakka dishes and Hakka snacks. Compared to Chaozhou and Cantonese cuisine, the flavor of Hakka cuisine is a little bit heavier on taste. "Fatty, salty and cooked," the characteristics are related with Hakka living environment and habits.

Shaoguan Hakka cuisine is slightly different from Dongjiang Hakka cuisine. The latter advocates authentic flavor of food while Shaoguan adds the spicy element to the cuisine. The reason for this is because the climate in Shaoguan is slightly colder, especially in winter, there is snowfall almost every year. Hakka people rely on chili to keep warm. Delicacies of Yao Village are also an important part of Shaoguan cuisine.

韶关有2000多年的历史，是客家人聚集地。这里淳朴的民风、独特的民俗、丰富的食材，造就了许多独具地方特色的美味佳肴。客家饮食可以分为两大类：一为客家菜肴，另一为客家小吃。客家菜与潮州菜、广府菜相比，口味偏重——“肥、咸、熟”，这与客家人居住的环境、生活水平和习惯相关。

韶关客家人的饮食与东江客家菜系又略有不同，东江菜系偏向“原汁原味，可口可心”，而韶关增加了“辣”的元素。这是因为韶关气候偏寒，特别是冬天，几乎年年都有降雪，劳动的客家人便要多吃辣椒以取暖。瑶寨美食也是韶关美食重要的组成部分。

MEIZHOU

A Bite of 'Capital of Hakka in the World'

梅州

世界客都的美味之旅

Meizhou is located in the northeast of Guangdong Province, bordering Fujian Province in the northeast and Jiangxi Province in the northwest. After several subsequent shifts of jurisdiction, it became Meizhou City in 1988. It comprises Meijiang District, Xingning City and six counties, namely, Mei County, Pingyuan County, Jiaoling County, Dabu County, Fengshun County and Wuhua County. It has an area of 15,864.5 square kilometers (urban area of 47.82 square kilometers). The minerals of the region include perlite, pyrophyllite, vanadium-titanium magnetite anddiabase commonly used as finishes. Meizhou local delicacies include Golden pomelo, tea and navel orange.

梅州市位于广东省东北部，与闽、赣交界。1988年设立地级市，辖梅江区、梅县区、平远县、蕉岭县、大埔县、丰顺县、五华县，代管兴宁市。行政区域面积15 864.5平方公里（其中市区面积47.82平方公里）。具有区域特色的矿产有珍珠岩、叶蜡石、钒钛磁铁及饰面用辉绿岩等。土特产有金柚、茶叶、慈（脐）橙等。梅州是中国历史文化名城，全球最具代表性的客家人聚居地，誉为“世界客都”，是叶剑英元帅的故乡、著名革命老区、海峡两岸交流基地，也是中国著名的文化之乡、华侨之乡、足球之乡、山歌之乡、金柚之乡、客家菜之乡、单丛茶之乡、油茶之乡。

Meizhou is a city with deep historical and cultural roots. It is the world's most representative Hakka settlement, known as the World Hakkas capital. It is hometown of Marshal Ye Jianying. It is a famous revolutionary base, and a cross-strait exchanges base. It has been crowned with titles such as Homeland of culture, Homeland of overseas Chinese, city of football, a land of folk songs and Homeland of Golden pomelo, Hakka cuisine, DanChong tea and Oiltea camellia. The main tourist spots are: Lingguang Temple; Ye Jianying Memorial Park; Hakka Museumof China; Renjinglu, the Former Residence of Huang Zunxian; Qianfo Temple (Thousand-Buddha Pagoda); Pankeng historical village; Hakka Tourism Park; Yearning Tea Plantation Tourist Attraction; Yanming Lake Tourist and Vocation Village; YinnaMountain Astronomical Park; Wangshou Mountain in Songyuan County; Heshui Reservoir in Xingning County; Shenguang Mountain; Yitang Reservoir in Wuhua County; Tanghu Thermal Mineral Mud-Bath Center; Nantai Mountain in Pingyuan County; Five-Finger Crags; Changtan Resort inJiaoling County; former residence of QiuFengjia; JiaolingShijia (Buddha) Cultural Centre; Dapu Fengxi Nature Reserve; Sanheba Martyr's Monument; The Sun Yat-sen Memorial Hall in Sanhe County; Tangkeng SPA in Fengshun County; Longgui waterfalls; Longjing River Rafting Scenic Area and Hakka round house, known as one of the five Chinese folk houses. In addition, Meizhou has distinctive Hakka folk customs, culture, arts and cuisines.

梅州主要旅游景区：梅江中国客家博物馆、黄遵宪故居人境庐、千佛塔、泮坑、客天下旅游产业园，梅县千年古刹阴那山灵光寺、叶剑英纪念园、雁南飞、雁鸣湖旅游度假村、阴那山天文科普园、松源王寿山，兴宁合水水库、神光山，五华益塘水库、热矿泥山庄，平远南台山、五指石，蕉岭长潭胜景、丘逢甲故居、释迦文化中心，大埔丰溪自然保护区、“八一”起义军三河坝战役烈士纪念园、三河中山纪念堂，丰顺汤坑温泉、龙归飞瀑、龙鲸河漂流以及中国五大民居之一的客家围龙屋等特色景点；还有独特的客家民俗风情、文化艺术，别具一格的风味饮食等。

YUNFU

Famous Cantonese Dim Sums in the "Hometown of Guangdong Folk Stone Art"

云浮

来自"石乡"的名菜名点

Yunfu City locates in the mid-center of Guangdong Province. It is known as the "Hometown of Guangdong Folk Stone Art." There are over 4,000 companies dealing with stone construction materials in the city, which help the city win the title of "Chinese Stone Construction Material Center," "The Land of Chinese Artificial Stone" and "Town of China Folk Culture (Stone Carving)." Yunfu New Port is the largest inland port in Guangdong. The main tourist attractions include: The Sixth Patriarch Hometown Tourist Resort, Golden Shuitai Hot Springs Area, Yunfu Beaulieu Hole and Luoding Dragon Bay Eco-tourism Area. Yunfu has been recognized as the city with splendid resources, heavenly treasures, outstanding people and smart land. The Nanjiang Culture, the Sixth Patriarch of Zen, and the ancient art of stone-cutting have long history and profound cultural background. Yunfu has been the birthplace of a number of outstanding figures, such as Master Monk Huineng (638-713), the Sixth Patriarch of Zen and founder of Zen's Southern Sect; Deng Fa, the famous leader of workers' movement in the 20th century and the famous anti-Japanese generals CaiTingkai. Yunan County folk dance Helou Dance is a national nonmaterial cultural heritage.

云浮市位于广东省中西部，素有“石乡”之称，全市有4 000多家石料建材及石料工艺企业，是中国石材基地中心、中国石材流通示范基地、中国人造石之都、中国民间文化（石雕）艺术之乡。云浮新港是广东内河第一大港。主要旅游景点有六祖故里旅游度假区、金水台温泉景区、蟠龙洞、罗定龙湾生态旅游区等。云浮市人杰地灵，南江文化、禅宗六祖文化和石艺文化历史积淀深厚，曾孕育出中国佛教禅宗六祖惠能、20世纪著名工人运动领袖邓发、著名抗日将领蔡廷锴等一批杰出人物。郁南县民间舞蹈“禾楼舞”是国家级非物质文化遗产。

Yunfu cuisine falls into category of Cantonese cuisine. With rice as a staple food, there are a variety of snacks made from rice. The food here basically follows the characteristics of Cantonese cuisine, emphasizing limitless range of ingredients. However, the flavors are slightly stronger, which is slightly different from the sweet flavor of the Pearl River Delta region. And diversity of dishes is also a feature of Yunfu diet. Based on a multiple choices of ingredients, many new dishes derive from traditional produce and food law. Yunfu City is Guangdong's main Broiler chickens production base. The famous restaurants here each have their own signature chicken dish, such as Yunfu Chicken in Pig Stomach, Luoding Tempeh Chicken, Yacan Deep-fried Crispy Chicken, Gongcheng Salt-baked Chicken in Xinxing City and Yunan Ilicifolius Chicken.

In recent years, Yunfu City has promoted the development of Yunfu food culture through organizing food festivals, and has named "Top 10 Dishes" and "Top 10 Snacks" of Yunfu City.

云浮饮食主要以广府菜为主，以米饭作为主食，或以米制成各式点心、小吃。基本遵循广府菜的特点：第一是选料广泛，无所不食。第二是博采众长，讲究实际。做菜时用盐稍多，口味偏重，跟珠三角地区用糖偏多，口味偏甜有细微区别。第三是菜式丰富，追求享受。凭着选料的优势，不少菜式从传统的搭配和食法中转换产生。云浮是广东肉鸡饲养主要基地，各地有名的酒家都有自己的招牌鸡，如云城猪肚包鸡、罗定豆豉鸡和亚灿吊烧鸡、新兴共成盐焗鸡、郁南大湾勒鼓鸡等，各具特色。

近几年来，云浮市通过举办美食节，推动了云浮美食文化发展，评出了云浮市“十大名菜”和“十大名点”。

壽

FOOD TRADITIONS

传统食俗

TAOTAOJU RESTAURANT

陶陶居

Taotaoju was famous for its Chinese pastries, moon cakes and dishes. It is renowned among overseas Chinese in Southeast Asia. The signature dishes include: Swine brain and fish head soup, Colorful shelled shrimp, Crab cooked with ginger and scallion, deep fried fish cut in the shape of a chrysanthemum flower, Salt roasted chicken, Roast goose, Yunnan ham with tripe pieces, and ham cooked with shrimp and crab meat. Celebrity chefs worked here include "King of Moon cake" Chen Dahui, "Flowery Qiang" Cui Qiang, Liao Bai, Liao Guoxi, etc. Moon cake with cured ham and five nuts -- almond, pistachio, pecan, cashew and macadamia, and Bailingsu with five-nuts stuffing are two must-try pastries in Taotaoju Restaurant.

Taotaoju Restaurant was founded in the Qing Dynasty. It is one of the oldest restaurants in Guangzhou. Located in No.20 Di Shi Fu Road of Liwan District, the place was anacademy of a rich man. The meaning of the name Taotao means sipping tea with a merry mood. After several renovations, in addition to its antique beauty, the façade of the building becomes more magnificent. The columns off the road were carved with the antithetical couplets written by contemporary calligrapher Qin Esheng. The vividly carved relief also added the antiquated elegance to the building. In the 1990s, many painting and calligraphy exhibitions were held in the Shuanghua Court on the third floor of Taotaoju, which demonstrates the traditional cultural tastes of the building. Cantonese opera performance has been a tradition in Taotaoju Restaurant. BaiJurong, the Cantonese opera actor who got his fame in playing young male characters, used to perform here. His performance had attracted a lot of literati and ordinary people.

陶陶居是广州最古老的茶楼之一，创于清朝末年，店址在广州市荔湾区第十甫20号，原是大户人家的书院。陶陶居的寓意是来此品茗乐也陶陶。如今，几经装修后的陶陶居，门面更加富丽堂皇，古色古香，马路边的柱上刻着当代书法家秦咢生手书民国时征联的典雅长联，还有栩栩如生的浮雕显出这老字号的大家气派。20世纪90年代，三楼的霜华小苑还连续举办书画展，体现了传统的文化品位。陶陶居一直开设有广东曲艺茶座，粤剧著名的“小生王”白驹荣当年常到此献艺，因此招徕了不少文人雅士和市井小民。

陶陶居以经营中式饼食、中秋月饼和茶面酒菜驰名，在东南亚一带华侨中享有盛誉。名菜有猪脑鱼羹、五彩鲜虾仁、姜葱炒肉蟹、西湖菊花鱼、手撕盐焗鸡、片皮挂炉鸭、云腿爽肚、雪里藏珍等。点心名师先后有号称“月饼泰斗”的陈大惠、绰号“百花强”的崔强，以及廖柏、廖国喜等，五仁咸肉月饼及五仁白绫酥就是陶陶居脍炙人口的点心。

LIANXIANGLOU RESTAURANT

莲香楼

In 1889, in the bustling West area of Guangzhou, a pastry shop started operation. The pastry shop is the predecessor of the time-honored brand Lianxianglou. At that time, a pastry cook surnamed Chen used lotus seeds to make pastry stuffing. The unique taste has attracted a lot of customers.

In the Guangxu years in Qing Dynasty, the pastry shop was renamed Lianxianglou. In order to ensure quality of its pastries, the shop used only the lotus seeds of the year from Hunan Province. Because of its superb quality, the business thrived. In 1910, Chen Ruyue, anImperial Hanlin Academy bachelor, suggested to change the character of the shop name "Lian" into the Chinese character of Lotus after tasting the bakery. He handwrote the shop mark of "Lianxianglou (Lotus House) Restaurant" for the restaurant. The logo is still in use today. Since then, the bakeries with lotus seeds stuffing produced by Lianxianglou Restaurant has attracted more and more people. The restaurant hence gained the fame of "No.1 Lotus Seed Stuffing Bakery." The business of Lianxianglou Restaurant thrived during the past century, and it has developed into a food chain dealing with restaurants, pastry, trade and lotus seeds stuffing production business.

Lianxianglou works hard in developing moon cakes. It produces more than 40 varieties of moon cakes, among which over 20 were made with lotus seeds. Each Mid-Autumn Festival, its moon cakes were packed and exported to Europe and the United States. In 1993, Lianxianglou was granted "China Time-honored Brand" by the Ministry of Internal Trade, and was named as one of the National Super Restaurants in 1998 By the Bureau of Internal Trade.

1889年，在广州繁华的西关一带，一间专营糕点美食的糕酥馆开业了，这就是广州老牌糕饼店莲香楼的前身。老制饼陈师傅用莲子来制作饼点馅料，使之独具一格，吸引客人。

光绪年间，糕酥馆改名为“连香楼”。为了保证饼食质量，该店严格选用当年产的湖南湘莲。由于制作讲究，生意日渐兴隆。1910年，一位名叫陈如岳的翰林学士，品尝莲蓉食品后，提议将连香楼的“连”改为“莲”，并手书“莲香楼”店招。三个字雄浑大气，至今沿用。从此，莲香楼制作的莲蓉食品，进入千家万户，被誉为“莲蓉第一家”。近百年来，莲香楼生意始终兴旺。今日的莲香楼，已发展为集餐饮、饼食、贸易、蓉馅生产于一体的连锁型企业。

莲香楼大力发展名牌月饼的生产，月饼品种达40多个，以莲蓉为主料的月饼品种便有20多个。每年中秋节，莲香月饼就远渡重洋，销往欧美等地。1993年莲香楼被国内贸易部授予“中华老字号”称号，1998年经国内贸易局批准为“国家特级酒家”。

PANXI RESTAURANT

泮溪酒家

Panxi Restaurant, dubbed as "National Super Restaurant", "China's Famous Restaurant" and "China's Time-honored Brand", is located near the beautiful Liwanhu Lake Park, formerly the imperial garden for the king of Nanhan Dynasty more than 1000 years ago. In 1947, Li Wenlun, a Cantonese, founded a small wine shop full of rural charm, and named it Panxi Restaurant after a little brook nearby. Designed by Mo Bozhi, China's renowned landscape architect, the restaurant is composed of rockeries, fish pond, zigzag corridors, a mid-lake restaurant and a seafood restaurant, all built and decorated with Lingnan style.

The restaurant boasts a team of top-notch bakers and chefs. In the 1960s, Luo Kun, a renowned dim sum master, introduced the idea of a "dim sum dinner", or Chinese Pastry feast, which includes all kinds of dim sums and adorn them with various exquisite decorations.

In 1983, eight dim sums,including rabbit-shaped dumplings in green grass, peach-shaped sweet pie, puff pastry with quail eggs, beehive-shaped yolk pie, fried soup dumplings, crystal prawn pie, Pantang horseshoe-shaped cake and fragrant apple pie, have been honored as "Famous Chinese Pastries" by the Guangzhou Municipal Government. Nowadays, in addition to Cantonese cuisine, the restaurant also serves Sichuan, Shandong and Jiangsu Cuisine.

泮溪酒家是国家特级酒家、中华餐饮名店、中华老字号，毗邻风光旖旎的荔湾湖公园。这里是1 000多年前南汉王的御花园故地。1947年，粤人李文伦在这里创办了一家充满乡野风情的小酒家，并以旁边的小溪“泮溪”命名。酒家由中国著名园林建筑师莫伯治设计，荟萃了岭南庭园特色和装饰艺术，由假山鱼池、曲廊、湖心半岛餐厅、海鲜舫等组成，其布局错落有致，加上荔湾湖的衬托，更显得风景如画。

泮溪酒家拥有一支全国最佳点心师和特级厨师团队，20世纪60年代，以罗坤为主的师傅运用食品拼盘的方法，给点心伴以象形的图案和花边，使点心登上筵席，成为名副其实的“点心筵席”。

1983年，八大名点（绿茵白兔饺、像生雪梨果、鹌鹑千层酥、蜂巢蛋黄角、生炸灌汤包、晶莹明虾脯、泮塘马蹄糕、清香苹叶角）被广州市人民政府命名。现在的泮溪酒家，除了粤菜，还增加了川、鲁、淮扬等菜式，成为集饮食大成的美食“大观园”。

GUANGFU TEMPLE FAIR

广府庙会

Guangfu Temple Fair, a traditional folk activity in Guangdong Province, has been newly upgraded by the city's Yuexiu District and become a one-week carnival held annually from the 15th day to the 21st day of the first lunar month on Chinese Calendar. Every year, the temple fair focuses on one theme. The theme of the 2015 event is "Get Together and Share the Happiness in Guangfu Temple Fair", which includes 10 activities like puppet performance, Chinese unique folk arts, martial arts performance and lantern show. It covers several elements of Chinese culture like blessings, folk customs, food and entertainment.

Many unique activities of Guangfu traditions and culture can, again, find their way to the public arena through temple fairs. The most visited part of the temple fair is the food section where many teahouses of unique style and time-honored brands would display their specialties to the visitors. The fair not only showcases famous local snacks but also traditional gourmet food that is further developed by Hong Kong and Macau. In 2015, the organizing committee of the temple fair, in cooperation with renowned food companies along the core culture zone in Beijing Road, developed an O2O business model for Guangfu cuisine that greatly promoted their growth.

广府庙会是广东省汉族传统民俗文化盛会，近年来由广州市越秀区重新打造，成为一项广府文化嘉年华活动。庙会选择在传统元宵节期间，即农历正月十五至廿一举办，为期一周。每年都有特定的主题，2015年的活动主题为“广府庙会，幸福相约”，涵盖木偶荟萃、中华绝活、武林大会、元宵灯会等10个主题活动，包含了祈福文化、民俗文化、美食文化、商贸休闲文化等丰富的内容。

通过庙会，许多广府民俗文化特色活动得以再现。其中人群光顾最多的就是庙会上的美食专区，众多广府味道特色酒家及老字号参展，展位与固定店铺互相补充，让吃货们体验到更精彩的美食历程。这里不但有广州知名的地道小吃和精致单品，还有香港、澳门发扬光大的广府传统美食。2015年还利用网络平台联动北京路核心文化区范围内的知名美食企业，打造广府美食O2O模式，带动广府美食创新升级。

BOLUODAN MILLENNIUM TEMPLE FAIR

千年庙会波罗诞

With a history of over a thousand years, Guangzhou's Boluodan Millennium Temple Fair is one of the most influential folklore events with unique traditions of Han nationality in the Pearl River Delta region, and is also China's only existing sacrificial ceremony dedicated to Mazu, Goddess of the Sea. It is held between the eleventh and the thirteenth day of the second lunar month on Chinese Calendar. The last day of the event, namely the thirteenth day of the second lunar month, is called Zhengdan, or Boluodan, meaning the birthday of Mazu.

During the fair, hundreds of thousands of visitors, including villagers in the Pearl River Delta Region as well as devout men and women, would come to the Nanhai Temple, or South Sea God Temple, for worshiping, sightseeing, shopping and enjoying delicious food. As a folk saying goes, visiting the Nanhai Temple is more important than marrying a wife, which demonstrates the significance of the temple fair.

Food festival during the temple fair is one of the most popular events. The food court, located in the front door of Nanhai Temple, not only serves gourmet food and delicious snacks in Guangdong, Hong Kong and Macau, but also provides delicious food from all over China like Hangzhou, Xinjiang, Dalian, Fujian, Shaanxi and Taiwan. Moreover, the food festival also features unique cuisine across the globe, like those from Argentina and Brazil, a real bonus for foodies.

In recent years, local government has stepped up application efforts for national intangible cultural heritage. On June, 2011, Boluodan traditional folk custom was included in the Third National Intangible Cultural Heritage List issued by the State Council.

广州波罗诞庙会已有千年历史，是珠江三角洲地区独具特色的汉族传统民俗节庆活动、最具影响力的民间庙会，也是现今全国唯一对海神进行祭祀的活动。每年农历二月十一至十三举行，其中十三为正诞，也叫菠萝诞，即南海神诞。

庙会期间，珠江三角洲一带的村民和善男信女便结伴来到南海神庙，祈福，观光，购物，饮食，人数达数十万。广州民间俗语有云“第一游波罗，第二娶老婆”，可见庙会影响之大。

庙会开设的美食节是最吸引人的活动之一。位于南海神庙正门的美食广场，不但汇聚了粤港澳的传统美食、特色小吃，还有来自杭州、新疆、大连、福建、陕西、台湾等地的地方特色美食，以及来自阿根廷、巴西等国的国际特色美食，让吃货们一饱口福。

近年来，当地政府启动了南海神庙申报国家级文物保护单位及“波罗诞”“波罗鸡”等申报国家非物质文化遗产工作。经过多方努力，2011年6月，国务院公布了第三批国家级非物质文化遗产名录，黄埔民间信俗波罗诞（民俗）榜上有名。

QIQIAO FESTIVAL

乞巧节

Guangzhou boasts the grandest celebration of Qiqiao Festival in China. It covers major procedures like worshiping the celestials, praying for wisdom from the Weaver Maid, reciting traditional prayers for dexterity in needlework, having fancy dinner known as the Weaver Maid's feast, and watching traditional operas known as the Weaver Maid's performance. The festival focuses on worshiping the celestials for cleverness in needlework and marital bliss, which calls for handicraft making including Qixi mascots and Magpie Bridge in miniature. Guangzhou's unique worshiping procedure in the Qiqiao Festival are not only toward the Altair and Vega stars, known respectively in Chinese folklore as the Cowherd and Weaver Maid, but also the latter's six sisters, which greatly demonstrates the local custom of gods worshiping.

People in Guangzhou have long attached much importance to celebrating this festival. As Liu Kezhuang, a poem in Song Dynasty (960-1279 B.C.) wrote, Cantonese celebrate Qiqiao Festival all night long. From the late Qing Dynasty to the early Republic of China, the celebration was especially popular among the general public. There are a lot of folk songs dedicated to this theme and Guangzhou has gradually developed several street dedicated to selling needlework items.

Normally, married women are not allowed to participate in the Qiqiao Festival. However, newly-married women should take part in a ceremony in the first Qiqiao Festival the year after their marriage. During the ceremony, the newly-married should offer pears to gods, which means bidding farewell to maidenhood, and also offer sweet wine, red eggs and pickled ginger, which symbolizes productivity and fertility.

广州乞巧节隆重的仪式为国内他处所未有，节日活动含摆巧、拜仙、乞巧、吃七娘饭、看七娘戏等内容；节日以“摆巧”为具体内容，并形成独具特色的传统七夕工艺系列作品，如斋塔、芝麻香、鹊桥景观、七娘盘、七夕公仔等；独具特色的“拜仙”仪式是广州乞巧节的第三个重要特色，祭拜的对象不仅有牛郎、织女二星，还包括织女的六个姐妹，体现了该地区自古以来浓厚的神灵崇拜的风习。

广州人历来重视过此节，宋人刘克庄诗云：“粤人重巧夕，灯火到天明。”清末民初，此风尤为兴盛。大量的竹枝词反映了羊城乞巧节的盛况，广州还形成了上九甫、下九甫、第十甫、十六甫等专卖女子物品的乞巧街市。

广州七夕拜仙，已婚女子一般不能参加，但新婚后的新娘在过第一个七夕时，要举行一次“辞仙”仪式，即在初六晚上祀神时，除了牲醴、红蛋、酸姜等（取兆得子），还要加上雪梨或沙梨，表示与姑娘节离别之意。

DRAGON BOAT RACE

端午龙舟

Dragon Boat Racing is a cultural sport event which enjoys a long history in the Pearl River Delta Region. In late April every year, every town in the region would pick an auspicious day to give sacrifice in the temple and then the eldest man in the community would lead other young man to get out of the dragon boat which was hidden in the dock last year. Later on, the most respected senior in the village will clean the boat and almost 80 strong villagers will begin practicing in before the race.

On the fifth day of the fifth month of China's Lunar Calendar, teams from every town will compete in the race. The winner will be awarded with spirits and roasted pick for dinner. A dragon boat feast will be served to the contestants and organizers of the event. After dinner and a sacrificial ceremony, the dragon boat will be buried into the original dragon boat dock. The general practice for preparing dragon boat feast is to select several excellent cooks from the village who will deliver a delicious meal with all kinds of meat dishes like fish, pork, chicken and duck. The names of those dishes are also very auspicious, like lucky roasted pork.

珠江三角洲河网交错，赛龙舟成为一项源远流长的传统民俗文化活动。每年的四月下旬，各乡选好吉日，在宗祠、庙宇拜祭龙舟头、龙舟尾后，由村内最年长的老人率众将上年埋藏在龙舟坞里的龙舟取出，洗净修补上油。然后由村中德高望重者用碌柚叶等为龙舟净身，在龙舟插上龙头龙尾，并在龙头龙尾上簪花戴红进行龙舟下水仪式。年轻力壮的近80名男子将龙舟划到河涌上演练，谓之“出龙”。到了五月初五，各乡的龙舟就会在河道上进行赛龙夺锦，胜出者奖励烧酒和大只烧猪。当晚，还要大摆“龙舟饭”，参加龙舟竞渡和为龙舟赛出力的人士都可以吃龙舟饭。吃完龙舟饭，一番拜祭仪式后，将龙舟重新埋到河涌的龙舟坞里，谓之“收龙”。龙舟饭传统的做法是，先从村民中挑选多名烹调高手做师傅，到各家各户借来炊具、碗碟，炒大锅菜。大鱼大肉是一定的，鸡鸭鱼一应俱全，名称也以吉祥喜庆的为主，如鸿运烧肉、红灼大沙虾、腰果锦上丁、绍菜扒鱼丸等。

TRADITIONAL WEDDING IN CHAOSHAN REGION

潮汕传统婚礼

Chaoshan people value much on wedding customs. In ancient times, they attached much importance to legally-binding marriage between the bride and groom who are introduced by a matchmaker. Though etiquette and customs of a wedding ceremony vary among different regions, it shares the following six procedures, namely proposing, birthday matching, presenting betrothal gift, presenting wedding gift, selecting the wedding date and the wedding ceremony. The last procedure, wedding ceremony means that the groom should pick up the bride from her family.

The procedures on wedding ceremony are complicated. Before leaving the bride's house and after arriving at the groom's house, the bride must have "five specially-made dishes", distribute money to her siblings, and have a dinner with his sisters. Before the bride sits onto the sedan chair, her family would usually spray some water in front of the chair. Moreover, one of her family member will dip a pomegranate flower into the water and spray it on the ground wherever the sedan covers. As soon as they entered the bridal chamber, the newly-weds will have some sweet glutinous rice dumplings. After finishing two rice dumplings, they would exchange their bowl and empty the rest of the dumplings. Moreover, the bride would also have five specially-made dishes once again, including swordfish, pork liver with scallions, fried pork with Chinese chives, meatballs, as well as glutinous rice or smashed sweet potatoes. There will be four cups full of brown sugar placed on four corners of the table. Whenever the bride picks up some food, she would dip it in the brown sugar, a custom that symbolizes a sweet marriage. Swordfish symbolizes able-mindedness, pork liver with scallion means cleverness, understandable and promotion in the workforce, Chinese chives means longevity and prosperity, meatball means getting together and sweet potatoes or glutionous rice means sweetness in marriage. All of the five dishes represent best wishes after their marriage.

潮汕结婚习俗颇为讲究，旧时注重“明媒正聘”。其礼俗虽各地繁简有异，但较一致的提法和仪礼是提亲、合婚、定亲、行聘、请期、迎亲，俗称“六礼”。迎亲（迎娶），即男方到女方家迎娶新娘。迎亲的程序繁多，仪式隆重。新娘在娘家和到夫家后都要分别“食五碗头”，出门前还要给兄弟姐妹“分钱米”，吃“姐妹桌”。新娘上轿前，家人要端一盆清水，轻洒向花轿，有的则提一小桶清水，用红花（即石榴花）蘸清水喷洒花轿和花轿经过的一段路程。新郎新娘进入洞房要吃“合房圆”。新郎新娘各吃两粒甜圆后互换圆盏，再吃两粒圆，俗称交杯换盏。进洞房后，新娘要吃“五碗头”：龙箭鱼、猪肝煮葱、韭菜炒猪肉、肉丸、甜碗——糯米饭或芋泥。桌上四角放四个甜杯，杯里放红糖，每夹一菜都蘸一点红糖，意取从头到尾件件甜。龙箭鱼象征能干，猪肝煮葱象征有官做、聪明、通情达理，韭菜象征长命和快发，肉丸象征团圆，甜碗取生活甜蜜，都属吉祥如意的美好愿望。

OPEN AIR BANQUET IN SHUNDE

顺德流水席

The traditional open air banquet is the reflection of Cantonese food culture and also the extension of Lingnan's unique culture of ancestral worship.

This open air banquet cannot possibly works out single-handedly but calls for teamwork from volunteers of the community. It can be compared to an assembly line that needs manpower to engage in organization of the feast, preparation of the food and cleanup of dishes.

There are eight Cantonese classic dishes and one soup in every round of the banquet. Nowadays, this open air banquet has added one more procedure for diners after they finished their meals. They have to clean all the oil stains on their napkins for the next round of diners. It is quite spectacular to see hundreds of diners washing napkins at the same time and this group behavior is said to symbolize happiness and auspiciousness for the coming new year.

传统的流水席是广东饮食文化的缩影，也是岭南祠堂文化的一种延伸。

流水席讲的是分工，这么大的宴席仅靠一个人是无法完成的，多需要街坊们一起来帮忙，从前期的准备到吃完后的清洗，就是一个完整的流水似的过程。

流水席每一轮都有八菜一汤，都是粤菜的经典菜式。现在的千人流水席还增加了特定安排——食客们就餐后须洗干净沾上油渍的餐巾，为下一轮的流水席食客提供方便。数百名食客一起洗干净餐巾的场面十分壮观，寓意街坊们齐齐“洗”迎新年。

LOCAL SPECIALTIES

缤纷特产

LYCHEE

荔枝

“荔枝”因其风味绝佳，深受民众喜爱，被誉为“果中之王”，唐代或更早即已列为贡品。杨贵妃最喜欢吃这岭南佳果，所以唐玄宗总要命人快马加鞭从岭南高州运送荔枝到长安。相比之下，宋代苏东坡的口福更好，他虽然被贬谪岭南，却因此有了“日啖荔枝三百颗”的经历。荔枝不易保鲜，一般离枝即食，长途运输之后色香味就损了很多。

广东荔枝品种繁多，主要有三月红、水东、妃子笑、桂味、糯米糍、淮枝等，尤以“增城挂绿”最为珍稀，被称为“荔枝之王”。

广东各地多种植荔枝，茂名地区是全国最大的水果生产基地，是世界最大的荔枝生产基地，种植荔枝面积176.57万亩。而惠来县是粤东最大的荔枝生产区，被国家林业局正式命名为“中国荔枝之乡”，如今在葵潭镇千秋镇村还保留着上千年树龄的荔枝，它们依然叶繁枝茂。

Lychee, dubbed as the king of fruit in Asia, is very popular among Chinese people due to its sweet flavor and fragrant smell. As early as Tang Dynasty (618-907), the fruit has become tribute to the royal family. As it was the favorite fruit of Emperor Li Longji’s favored concubine Yang Yuhuan (Yang Guifei), the emperor had the fruit delivered to her at great expense to the capital. Compared to Concubine Yang Yuhuan, Su Dongpo, the master poet in Song Dynasty (960-1279), was much luckier because he could enjoy “300” lychees a day as he was demoted to the Lingnan region, something of a blessing in disguise. Since its flavor is lost in the process of transportation, lychee is better eaten fresh once picked up from the tree.

Lychee has many different varieties, but the best one is from Zengcheng city where the “king of lychee” had been cultivated with the best flavor.

Lychees are extensively grown in China’s Guangdong Province, and Maoming prefecture boasts the largest fruit production base in China and the largest lychee production base in the world with a cultivation area of 1765,700 mu, or 117713.33 hectares. Huilai County, the largest lychee production area in eastern Guangdong, has been named “hometown of lychees”. Qianqiuzhen Village, Kuitan Town in Huilai County is home to lychee trees with a history of over a thousand years.

LONGAN

龙眼

Longan is dubbed as one of the four famous fruit in Lingnan region, along with lychee, banana and pineapple. The longan is so named because it resembles an eyeball when its fruit is shelled (the black seed shows through the translucent flesh like an iris). The fruit is also called Guiyuan, as it is harvested on August and its shape is round.

Cultivation of longan can be traced back to the Han Dynasty (206 B.C.- A.D. 220) of about two thousand years ago, and was first documented in the Book of Later Han. Longan is one of the most well-known and well-used medicinal fruits in Chinese medicine. It strengthens both the digestive system as well as the heart. This means that longan is especially good for blood deficiency symptoms that are related to mind or emotions. During the Han Dynasty, it has been widely used in Chinese Medicine. In the Ming Dynasty (1368-1644), the great medical doctor and herbalist Li Shizhen recognized the fruit as beneficial for people's health. It now enjoys wide application as one of the favored medicinal foods for insomnia, poor memory, stress, and palpitation.

龙眼，又称“桂圆”，与荔枝、香蕉、菠萝并称为“岭南四大名果”。龙眼去皮则剔透晶莹偏浆白，隐约可见肉里红黑色果核，极似眼珠，故名“龙眼”。其成熟期在农历八月，由于古时称八月为“桂”，加上龙眼果实呈圆形，所以又称桂圆。

龙眼栽培历史可追溯到2000多年前的汉代，文献记载最早见于《后汉书》。其味甘甜，属于温热性质，具有补益心脾、养血安神的功效。在汉朝，就已作为药用。明李时珍曾有“资益以龙眼为良”的评价。至今，龙眼仍然是一味补血安神的重要药物。

PINEAPPLE

菠萝

Pineapple is one of the four famous fruits in Lingnan region. It is very popular due to its sweet flavors and refreshing juicy texture.

Pineapple can be eaten fresh or in a processed form, notably canned ones which can preserve the original flavor of the fruit. It also finds application in culinary use like in the case of fried pork with pineapples and pineapple fried rice. Pineapples in those dishes facilitates digestion and makes them less greasy.

Xuwen County in Guangdong Province boasts a pineapple production base with its cultivation area reached 250,000 mu (16666.67 hectares). The pineapples produced here, notably the ones grown in Yugonglou, are big, juicy, crispy and sweet. The county produces more than 340,000 tons of pineapples every year, ranking the first in terms of total production in China. Its fresh and processed pineapples are popular at home and abroad. Canned pineapples now are exported to 60 countries and regions all over the world including Europe, America, the Middle East, Africa and the Oceania.

菠萝为“岭南四大名果”之一，以其肉色金黄、香味浓郁、甜酸适口、清脆多汁而深受民众的喜爱。

除鲜食外，菠萝多用以制罐头，因其能保持原来的风味，被誉为“国际性果品罐头”。此外，菠萝还适合做各种美味料理，能去除油腻，帮助消化。比如菠萝古老肉、五彩菠萝炒饭、菠萝船水果沙拉等。

广东湛江徐闻县有25万亩“菠萝的海”，一眼望去都是漫山遍野、无边无际的绿色海洋。此地生产的菠萝果大、果汁多、果肉脆、果味甜，以愚公楼菠萝为最佳。徐闻菠萝年产量34万多吨，是中国菠萝生产第一大县，鲜菠萝及其加工产品畅销海内外，尤其是菠萝罐头畅销欧美、中东、非洲和大洋洲等60多个国家和地区。

WAMPEE

黄皮

黄皮原产我国南部，已有1500多年的栽培历史，是广东人最熟悉的夏季水果之一。黄皮果色泽金黄、光洁耀目，根据性味，可分甜、酸两个系统，有些品种甜酸适口、汁液丰富而具香味，是色、香、味俱佳的水果，可与荔枝并称。

对广东人来说，黄皮一身是宝，不但生津止渴，还具有消食健胃的功能，所以广东民间有“饿吃荔枝，饱吃黄皮”的说法。此外，民间也喜欢拿水煎黄皮叶来防治感冒，或者用黄皮树根来治气痛。其果皮和果核都是入药良材，有利尿消肿、行气止痛等功效。

Wampee, with a cultivation history of 1500 years, is a highly-esteemed fruit in southern China, where sour, sub-acid, and sweet varieties are known. Some varieties have a sweet-sour flavor and juicy texture which can match lychee in terms of color, aroma and taste.

As a Cantonese proverb goes, “lychees can feed your hunger while wampees can help digestion”. Cantonese viewed the fruit as a God-sent treasure which has can help produce saliva, slake thirst and facilitate digestion. Moreover, boiled wampee leef can also be used to prevent colds and flu and the root of wampee trees can be used to cure pain in one’s body. Its skin and kernel, which can be act as diuretic and reduces swelling, is widely applied in Chinese Medicine. Other therapeutic qualities consist of coolant, stomachic, as well as antihelmintic.

CITRUS SHATANGJU

砂糖橘

This citrus variety is named after is origin Shatang. It is indigenous in Guangning and Sihui in Guangdong Province and is widely grown in Yunfu, Qingyuan, Deqing, Nansheng, Yunan and Fengkai. This top quality variety is a local specialty in Guangning, Sihui and Fengkai.

This yellow fruit is oblate whose skin is easy to peel off. The most famous and authentic ones are grown in Sihui, a city with a cultivation history of over a thousand years. Home to citrus Shatangju, tributes orange, sweet organge, Sihui boasts the top quality variety with the best flavor and taste.

砂糖橘，又名十月橘，因出产于砂糖故名。主产地有四会、广宁、云浮、清远、德庆、南盛、郁南、封开，为广宁、四会、封开传统土特产，是柑橘类的名优品种，被誉为“一枝独秀”。

砂糖橘果实扁圆形，色泽橙黄，皮薄易剥。尤以四会出产的为正宗。四会柑橘种植历史有1000多年，是砂糖橘、贡柑、蜜橘的正宗原产地，所产的砂糖橘鲜美而极甜，细腻无渣，实为极品！

佳果 Fruit

GREEN PLUM

青梅

Green plum, a specialty in Puning of Guangdong Province, is cultivated since 700 years ago. It is also among the National Geographical Indication Products List. Plums cultivated in Puning are big with thick and juicy texture and would maintain the color and shape after dried. With a plantation area of 160,000 mu (10666.67 hectares), Puning boasts an annual production of 35,000 tons, of which 28,000 are sold overseas. The city ranks the first in terms of cultivated area, total production and exports among county-level cities all over China. It was honored as the hometown of green plum in April, 1994 by the Chinese Association of Agricultural Science Societies (CAASS).

Green plums can be processed into various kinds of "preserved fruit" such as candied plum, salty plum, crispy plum, lover's plum, lovesick plum, licorice plum and plum candy. The crispy plum, which is bright in color and crispy in texture and maintained the original flavor the fresh fruit, has been included in China's National New Product List.It is dubbed as "China's valuable preserved fruit"by foreign businessmen.

80 percent of the processed green plums are sold overseas to 10 countries and regions like Japan, Republic of Korea, Southeast Asia, Europe and North America. It contributes a large share of foreign exchange earnings in the city.

青梅是广东省普宁市特产，国家地理标志产品。种植历史已有700多年。普宁青梅以其果大、肉厚、核小、酸度高，果皮柔韧、不易破损、肉质柔软、晒干率高、色泽鲜艳、成品保色期长等优点而著称。目前全市青梅种植面积约16万亩，年产量3.5万吨，出口量2.8万吨，种植面积、总产量、出口量均居全国县（市）级首位，1995年4月被中国农学会特产经济专业委员会命名为“中国青梅之乡”。

普宁青梅加工制品品种丰富，主要有干湿梅、咸水梅，蜜饯类有蜜梅、酥梅、陈皮梅、情人梅、相思梅、甘草梅、话梅等，还有梅酱、梅酒、梅汁等三大系列几十个品种。其中酥梅因色泽雅观、肉质酥脆、品质上乘，具有原果风味而被评为“国家级新产品”，被外商誉为“凉果之珍品”。

普宁青梅制品80%以上出口，产品畅销日本、韩国、东南亚、欧美、俄罗斯以及港澳台地区，成为普宁出口创汇的拳头产品。

CHAOSHAN KONGFU TEA

潮汕工夫茶

Kongfu Tea, with a literal meaning of "making tea with effort", is the most popular drink in Guangdong's Chaoshan region. According to ancient documents, the tea can be traced back to the Ming Dynasty and is well-established in the Qing Dynasty. During the Mid Qing Dynasty, it becomes a popular drink in the Chaoshan region and quickly spread to the neighboring areas like the Southeast Asia. Kongfu Tea is an integral part of Chaoshan's traditional culture. However, the tea processing technique has been replaced by instant tea in this fast-paced society.

The making of Chaoshan Kongfu Tea includes several procedures like selecting high quality tea, choosing pure water, selecting the right kind of tea sets and brewing the tea in hot water. The brewing procedures is the most complicated one which includes preparing tea sets and tea leaves, brewing tea leaves, rinsing tea leaves, heating the teapot and tea cups and pouring the tea to the cups for later enjoyment. A very important part is to smell the aroma of the tea by inhaling the steam. After smelling, drink by taking small sips that allow to fully enjoy the taste, aroma and qualities of the tea. Normally Chaoshan people would use Oolong tea to make the drink.

Kongfu tea can be enjoyed in any circumstances and forms. It represents major part of China's tea culture with historical, cultural, aesthetic and entertainment value.

工夫茶以广东潮汕地区最为知名。根据文献记载，它在明代兴起，清代烹治日臻完善。清中期以后，已在潮汕地区蔚然成俗，甚至辐射周边地区，流传到东南亚各地。作为一种文化现象，工夫茶是潮汕传统文化的重要组成部分。只是现代社会生活的快节奏，使传统的潮汕工夫茶艺已逐渐被简化的冲泡程式所代替。

潮汕工夫茶的程式主要有择茶、择水、择器、烹治，其烹治之法为治器、纳茶、候汤、冲茶、刮沫、淋罐、烫杯、洒茶。八法演尽，接着就是品茶，先闻其香，和气细啜、三嗅杯底、瑞气圆融。它除了冲泡独特，以“乌龙”茶系（半发酵茶）为主要茶料也是其突出特色。

饮用工夫茶不受环境限制，形式多样，如独饮、对饮或三五成群共饮等，有着历史文化、美学、社会、家庭文化、健身益智诸价值。

GREEN TEA

绿茶

Green tea, made from the leaves of Camellia sinensis that have undergone minimal oxidation, is the most widely consumed beverage in China, the largest green tea producing country.

There are many varieties of green tea in Guangdong Province including Gulao tea, Guangbei Yinjian tea and Dayinhao tea. Dayezhong green tea, a recent variety developed by phoenix narcissus in Yunnan Province, is widely grown in Guangdong due to its refreshing and aromatic flavor despite several times of brewing.

绿茶是中国产量最多、饮用最为广泛的一种茶。它是将采摘来的鲜叶先经高温杀青，杀灭了各种氧化酶，保持茶叶绿色，然后经揉捻、干燥而制成。清汤绿叶是绿茶品质的共同特点。

广东绿茶品质繁多、产量大，有产于鹤山古劳区的古劳茶、产于乐昌广北茶场一带的广北银尖、产于英德的大银毫茶等10余种。目前，广东绿茶主要是以云南大叶种凤凰水仙为主要品种生产的大叶种绿茶，是最近10多年新兴的茶叶产品。其由于具有滋味浓醇爽口、回甘味好、刺激性强、耐冲泡等特点而受到消费者的欢迎。

HERBAL TEA

凉茶

Since ancient times, the Cantonese have been famous for drinking herbal tea. Herbal tea is not kind of tea, but medicine soup boiled down from Chinese herbal medicine. Herbal tea is not necessarily cold and it works better to drink it hot. The most representative is Bansha Herbal Tea.

Herbal tea preparation techniques, handed down from generation to generation in a family tradition, have had a history of hundreds of years. The earliest Guangdong herbal tea is Wong Lo Kat Herbal Tea, invented by Wong Chat Bong from Heshan City of Guangdong Province in 1828 during the Qing Dynasty. Later, Wong Lo Kat Herbal Tea went abroad and has turned popular among the Chinese all over the world for more than 170 years. It is indeed the pioneer drinks for people to clear heat, remove toxin and regain beauty.

There are 54 recipes derived from the following 18 herbal tea brands: Wong Lo Kat, Shangqingyin, Jianshengtang, Denglao, Baiyuanshan, Huangzhenlong, Xuqixiu, Chunhetang, Jinhulu, Xingqun, Runxintang, Shaxi, Lishi, Qingxintang, Xinglinchun, Baoqingtang, Fuqingtang and Huangfuxing. These recipes and the herbal tea culture have been widely recognized among the public. For hundreds of years, a forest of herbal tea shops in Guangdong, Hong Kong and Macao have constituted a unique landscape of the Lingnan culture.

Herbal tea was included in the first batch of national intangible cultural heritage announced by the State Council in 2006. Now, it has been sold all over China and in nearly twenty countries including the US, Canada, France, Britain, Italy, Germany, Australia and New Zealand.

自古以来，广东人喝凉茶就喝出了名。凉茶不是茶，而是中草药熬出来的药汤；凉茶也不一定凉，热着喝效果更好。最有代表性的就是斑砂凉茶。

凉茶配制技艺以家族世袭传承下来，已有数百年历史。历史最早的广东凉茶是由广东鹤山人王泽邦于1828年（清道光八年）始创的王老吉凉茶。后来，王老吉凉茶漂洋过海，在世界各地的华人中流行180多年，可算是清热、解毒、美颜饮品的先驱。

王老吉、上清饮、健生堂、邓老、白云山、黄振龙、徐其修、春和堂、金葫芦、星群、润心堂、沙溪、李氏、清心堂、杏林春、宝庆堂、福庆堂、黄福兴等18个凉茶品牌的54个配方及其所构成的凉茶文化得到了民众的广泛认可。数百年来，林立于广东、香港、澳门的凉茶铺，形成了一道岭南文化的独特风景线。

2006年国务院公布的首批国家级非物质文化遗产名录，凉茶赫然在列。目前，凉茶销售范围已覆盖全国及美国、加拿大、法国、英国、意大利、德国、澳大利亚、新西兰等近20个国家。

LEI CHA (GROUND TEA)

擂茶

Lei cha or ground tea is quite unique in the great and profound Chinese tea art. As the special food for Hakka people, it is one of the traditional cuisines among the Hakka in Guangdong Province and Guangxi Zhuang Autonomous Region. It can be taken not only as part of their regular diet, but also a delicacy to entertain guests.

Every kind of food culture is poetic expression for and enlightenment on the philosophy of life. The production of Lei cha and its flavors fully demonstrate Hakka people's inheritance of Chinese traditional culture. Lei cha is actually a mix of tea leaves, rice, sesame, soybeans, peanuts, salt and orange peel, sometimes with Chinese herbal medicine included. Such ingredients are ground with a mortar and pestle into powder. Easy to produce, the tea tastes pure with a strong fragrant smell. It is also affordable and brings a lot of benefits. Due to a varying mix of ingredients, Lei cha can help people quench their thirst, feel cool and refreshing, relieve summer heat, strengthen the spleen and warm the stomach, nourish themselves and prolong life.

在中国博大精深的茶艺中，擂茶是一枝独秀的奇葩。它是一种客家民族风味的特色食品，是广东、广西客家人的传统美食之一，既可日常使用，也是待客佳肴。

每一种饮食文化，无不是生活哲学的一种诗化、一种启迪。擂茶的制作与风味充分表现了客家人对中国古代传统文化之传承。“擂”茶的用具是擂持和擂钵，基本原料是茶叶、米、芝麻、黄豆、花生、盐及橘皮，有时也加些青草药，制作简便，茶味纯，香气浓，且因配料不同，分别具有解渴、清凉、消暑、健脾养胃、滋补长寿等效用，经济而又实惠。

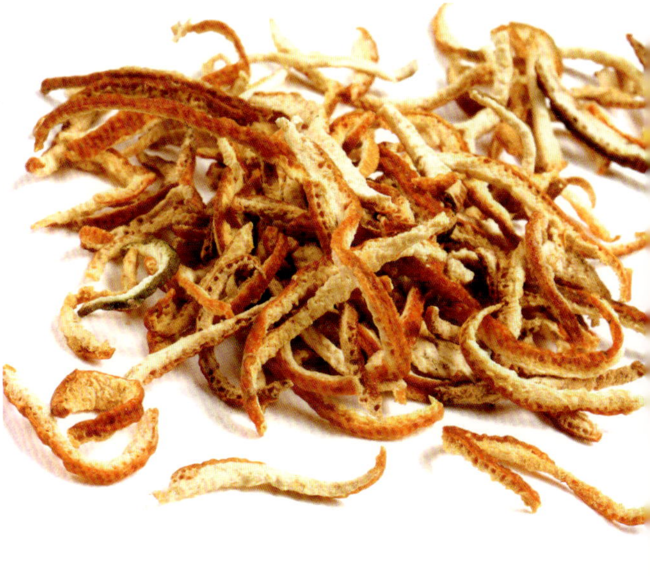

珠江啤酒
PEARL RIVER
珠江纯生 真正好纯生
第一瓶珠江纯生诞生于1997年，其经过全程密封灌装、低温膜过滤，使生啤酒的新鲜和营养得以保存。
PEARL RIVER
珠江啤酒
珠江
DRAFT BEER
纯生啤酒
精品
TM
生
珠江精品纯生啤酒
PEARL RIVER
珠江啤酒
珠江
纯生啤酒
DRAFT BEER
精品
TM
净含量:500mL
生

广州珠江啤酒集团有限公司

广州珠江啤酒集团有限公司成立于1985年，是一家以啤酒业为主体、以啤酒配套和相关产业为辅助的大型现代化企业，啤酒产能230万吨，目前拥有20家下属企业，是我国酿酒行业十强、国家环境友好企业和广东省高新技术企业。

珠江啤酒集团主要特点：一是技术创新。拥有国家级企业技术中心和博士后科研工作站，20项技术填补国内空白，率先研制推出我国第一瓶纯生啤酒和白啤酒，成为中国首个纯生啤酒示范生产基地。二是品质优秀。相继通过ISO9000、ISO2000、ISO14000、HACCP、EHS等国际体系认证，获首批“广州市政府质量奖”。三是环保领先。多项消耗指标位居世界先进行列，烟囱实现无烟排放、污水净化后可以养鱼，其中沼气回收发电、余热用于制冷的项目成为我国啤酒行业第一个入选联合国清洁发展机制（CDM）项目。

地址：广东省广州市新港东路磨碟沙大街118号　　电话：020－84206636
传真：020－84202502　　网址：www.zhujiangbeer.com

Guangzhou Zhujiang Brewery Group, Co., Ltd

Established in 1985, Guangzhou Zhujiang Brewery Group, Co., Ltd is a large state-owned enterprise taking beer industry as its main business and packaging and other relative industries as its supporting businesses. Owning 20 subordinate companies, it currently has a beer production capacity of 23 million. It is ranked the top ten in China’s brewing industry, and also awarded as national environment-friendly enterprise and new high-tech enterprise in Guangdong Province.

Zhujiang Brewery Group has three main features as follows: 1) technology innovation It has national enterprise technology center and Post-Doctoral Science and Research Workstation where 20 items of techniques were invented to fill the vacancy of domestic beer industry. It also takes the lead in developing and launches bottled draft beer and white beer in China, making it the first demonstrative production base for draft beer. 2) excellent quality It has successively passed the international certification system including ISO9000, ISO2000, ISO14000, HACCP, and EHS. It is among the first batch of enterprises to be awarded as “Guangzhou Municipal Government Quality Prize”. 3) advanced environment protection Multiple consumption indexes reached the world’s advanced level. For example, the chimney realizes smoke-free emission and purified wasted waster can be used to increase fish. It is the first brewery in China with its projects of recovering marsh gas collected from waste water treatment to generate electricity and using waste heat to refrigerate and to be selected into United Nations’ Clean Development Mechanism (CDM).

Address: No. 118, Modiesha Street, Xingang East Road, Guangzhou, Guangdong Province
Tel: 020－84206636　　Fax: 020－84202502
Website: www.zhujiangbeer.com

皇冠干邑

DOUBL

Co

百年洋酒品牌区域合作

皇冠酒庄

皇冠钻石红酒

皇冠金钻干邑

CROWN

ac

王者享受

赢未来

英国皇冠洋酒有限公司（CROWN WINE CO.,LIMITED UK），成立于1991年，总部设在英国伦敦，是一家经营二十多年的经验烈性酒专业公司。公司在2006年发展中国市场，带着专业的经营理念和丰富经验，致力把世界各国质优价美的好酒带给中国消费者。

英国皇冠洋酒有限公司属下的产品有：皇冠干邑系列、皇冠钻石酒系列、伏特加、朗姆酒、金酒、威士忌、龙舌兰等等，特别是来自法国干邑区的皇冠干邑系列酒深受中国市场的欢迎。皇冠干邑系列包括：皇冠干邑60年金伯爵、皇冠干邑50年银伯爵、皇冠干邑35年金钻、皇冠干邑25年银钻、皇冠干邑16年明珠、皇冠干邑VSOP、皇冠干邑海洋之星。其中皇冠干邑35年金钻多次在国际权威性比赛中屡获殊荣，它获得2010年国际酿酒技术金奖、2011年亚洲市场最受欢迎的烈性酒，冠军金奖！2011年英国伦敦国际烈酒挑战赛，金奖！并为广东省高尔夫球协会冠军奖品殊荣、爱斯克菲国际美食会唯一指定用酒、粤港澳唯一指定用酒和唯一指定供应商、世界环球夫人唯一指定用酒、世界华裔小姐指定用酒。皇冠钻石酒系列包括：大钻石阿马罗尼红葡萄酒、小钻石红葡萄酒、钻石白葡萄酒、六钻香槟起泡酒。百年洋酒品牌诚招全国区域合作伙伴。

CROWN WINE CO., LIMITED UK, established in 1991, headquartering in London, is a business with twenty-two year experience in experienced spirits. In 2006, adhering to the professional management philosophy and rich experience, the Company is committed to bringing the high quality and inexpensive worldwide wine to Chinese customers.

The products from CROWN WINE CO., LIMITED UK, namely are Crown cognacs, Crown Diamond wine series, vodka, rum, gin, whiskey, tequila, etc., among which, Cognac Crown series, are particularly popular in Chinese market. Cognac Diamond Crown of 35 years have won many titles in the international authoritative competitions, like the gold award in 2010 International Liquor-Making Technique Competition, the most popular spirits in Asian market in 2011 and the laurel of Guangdong Golf Association Champion. It also serves as the only designated wine in Global Universal Mrs. And the only supplies for Guangdong, Hong Kong and Macao. This hundred-year wine brand sincerely welcomes the regional partners in China.

免费加盟指导　零装修 零风险　店面形象统一

英国皇冠洋酒有限公司
广州保税区广保大道18段106-107铺
手机：13500012833（张生）　电话：020-34321819　传真：020-34321545
邮箱：doublecrowncognac@hotmail.co.uk
网址：www.doublecrown-cognac.com

廣中外名酒商城

5000 m²海珠酒城实景图

After thirteen-year development, Guangdong Chinese Foreign Vintage Wine Chain Co., Ltd , a trustworthy company, wins successively eight assured wine model enterprise titles in Guangdong and serves as Guangdong Alcohol Industry Association Vice Director-General Unit. Its two direct-sale stores (Haizhu and Panyu Shopping Malls), covering the area of 10,000 square meters, opened the e-commerce platform, that is the wide Chinese and Foreign Vintage Wine online store (www.gzwmj. com) and wechat mall (wechat No: egzwmj), which enables it to realize the online and offline (O2O) combination. Wine products operated by this company include renowned white spirit and wine at home and abroad. This company strives for the best cost-performance wine mall.

Guangdong Chinese Foreign Vintage Wine Chain Co., Ltd is launching a maker brand and looks for marketing partner. The signed partners can take part in sales and enjoy the seven in ten percent of profits. This philosophy of this platform is to offer the platform to help you start a business, form a partnership and leave the risk on me.

燕塘®

广东燕塘乳业股份有限公司由广东省燕塘投资有限公司、广东省粤垦投资有限公司、广东省湛江农垦集团公司和广东中科白云创业投资有限公司等股东发起设立，是农业部广东农垦的直属企业。2014 年 12 月，燕塘乳业成功登陆 A 股资本市场，在深圳市证券交易所挂牌上市（股票代码：002732）。作为目前广东本土乳制品行业唯一一家上市公司，燕塘乳业经过多年的发展，至今已成为集牧场、研发、加工、销售于一体，华南地区规模最大、设备最先进的乳制品生产企业之一。

Guangdong Yantang Dairy Co., Ltd is founded by Guangdong Yantang Investment Co., Ltd, Guangdong Yueken Investment Co., Ltd and other stakeholders together. It is the enterprise directly under the Guangdong Agricultural Reclamation of Agriculture Minister. In 2014, Yantang Dairy successively landed the A-share capital market and was listed on the stock market (stock code: 002732). As the only listed dairy company in Guangdong, Yantang Dairy, integrating pasture, development, processing and sales, has become one dairy products enterprises with the largest scale and the most advanced equipments in South China.

STRATEGIC LAYOUT

燕塘乳业秉承“市场导向，科技领先，质量第一，顾客至上”

Yantang Dairy adheres to the market-oriented, technology-leaded, quality first and customers highest.

CORPORATE DEVELOPMENT

精耕广东、放眼华南、迈向全国 始终为消费者提供新鲜、安全、营养的好牛奶

Yantang Dairy focuses on Guangdong, looks out onto South China and marches toward China. It always provides the customers the fresh, safe and nutritious milk.

广东燕塘乳业股份有限公司
广州市天河区沙河燕塘
服务热线：400-888-1956
网址：http://www.yantangmilk.com

采用独有发酵技术
并结合现代先进科技
采用低温发酵100%纯酿技术
保证100%无添加，100%桑葚，100%纯度的桑葚红酒
完全保留了桑葚丰富的营养成分及药用疗效
GUO YUAN FRUIT WINE
果园红桑果酒
中国长寿之乡养生酒
继啤酒，白酒，红酒之后
又一大创新美酒
VIP
350ML/218元
VIP
750ML/380元
VIP
礼盒/558元

南方日报出版社
NANFANG DAILY PRESS

出 版 人：周洪威

策划统筹：陈微尘

责任编辑：阮清钰　　黎　衡

责任技编：王　兰

责任校对：阮昌汉

书籍设计：书窗设计工作室

广告代理：广州市巨鑫文化传播有限公司

广州点房云广告策划有限公司

Publisher：Zhou Hongwei

Director：Chen Weichen

Editor：Ruan Qingyu Li Heng

Copy-editor：Wang Lan

Proofreader：Ruan Changhan

Design：Shuchuang Design Room

Adcraft：Guangzhou Juxin Cultural Communications Co.,Ltd

Guangzhou Dianfangyun Advertising Planning Co.,Ltd